An Introduction to Syntropy

Ulisse Di Corpo

Antonella Vannini

www.sintropia.it

CONTENTS

SYNTROPY

The notion of energy comes from the fact that physical systems possess a quantity that can be turned into a force.

This quantity can take the form of heat, mass, electromagnetism, potential, kinetic, nuclear and chemical energy.

Despite the fact that it is used and studied *"it is important to realize that in physics today we have no knowledge of what energy is."*[1]

The energy-mass relation:

$$E = mc^2$$

that we all associate with Einstein, was first published by Oliver Heaviside in 1890[2], then by Henri Poincaré in 1900[3] and by Olinto De

---

[1] Feynman R (1965), *The Feynman Lectures on Physics*, California Institute of Technology, 1965, 3.
[2] Auffray J.P., *Dual origin of E=mc2*:http://arxiv.org/pdf/physics/0608289.pdf
[3] Poincaré H,. *Arch. néerland. sci.* 2, 5, 252-278 (1900).

Pretto in 1904[4]. Olinto De Pretto presented it at the *Reale Istituto Veneto di Scienze* in an essay with a preface by the astronomer and senator Giovanni Schiaparelli.

It seems that this equation has come to Einstein through his father Hermann who was responsible for the lighting systems in Verona and who, as director of the "*Privilegiata Impresa Elettrica Einstein*", had frequent contacts with the Fonderia De Pretto that produced the turbines for electricity.

However, the $E=mc^2$ does not take into account the momentum, which is also a form of energy and in 1905 Einstein added the momentum (p), thus obtaining the energy-momentum-mass equation:

$$E^2 = m^2c^4 + p^2c^2$$

Since energy is squared (E^2) and in the momentum (p) there is time a square root is used and there are two solutions: negative time energy and positive time energy.

---

[4]De Pretto O., *Lettere ed Arti*, LXIII, II, 439-500 (1904), Reale Istituto Veneto di Scienze.

 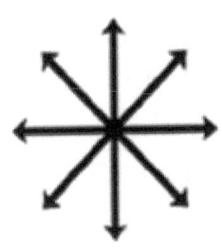

E^{-t}, *negative time energy, manifests as converging energy*

E^{+t}, *positive time energy, manifests as diverging energy*

Positive time energy implies causality, whereas negative time energy implies retrocausality: the future that acts back into the past. This was considered impossible and to solve this paradox Einstein removed the momentum, given the fact that it is practically equal to zero compared to the speed of light (c). In this way, we return to the $E=mc^2$.

However, in 1924 the spin of the electron was discovered. The spin is an angular momentum, a rotation of the electron on itself at a speed close to that of light. Since this speed is very high, the momentum cannot be considered equal to zero and in quantum mechanics the energy-momentum-mass equation must be used with its uncomfortable dual solution.

The first equation that combined relativity and quantum mechanics was formulated in 1926 by Oskar Klein and Walter Gordon and has two time solutions: advanced and delayed waves. Advanced waves were rejected, since they imply retrocausality which was considered impossible.

The second equation, formulated in 1928 by Paul Dirac, also has two time solutions: electrons and neg-electrons (now called positron). The existence of positrons was proved in 1932 by Carl Andersen.

Shortly after Wolfgang Pauli and Carl Gustav Jung formulated the theory of synchronicities. Starting from the dual time solution they came to the conclusion that reality is supercausal, with causes acting from the past and synchronicities acting from the future.

In 1933 Heisenberg, who had a strong charismatic personality and a leading position in the institutions and academia, declared the backward in time solution impossible. From that moment, anyone who ventures into the study of the backward in time solution is discredited, loses the academic position, the ability to publish and to talk at conferences.

Luigi Fantappiè studied pure mathematics at the Normale di Pisa, the most exclusive Italian University, where he had been classmate of Enrico Fermi. He was well known and appreciated among physicists to the point that in 1951 Oppenheimer invited him to become a member of the exclusive Institute for Advanced Study in Princeton and work directly with Einstein.

As a mathematician Fantappiè could not accept that Heisenberg had rejected half of the solutions of the fundamental equations and in 1941, while listing the properties of the forward and backward in time energy, Fantappiè discovered that the forward in time energy is governed by the law of *entropy*, whereas the backward in time energy is governed by a complementary law that he named *syntropy*, combining the Greek words *syn* which means converging and *tropos* which

5

means tendency.

Entropy is the tendency towards energy dissipation, the famous second law of thermodynamics, also known as the law of heat death. On the contrary, syntropy is the tendency towards energy concentration, increase in differentiation, complexity and structures. These are the mysterious properties of life!

In 1944 Fantappiè published the book *"Principi di una Teoria Unitaria del Mondo Fisico e Biologico"* (Principles of a Unitary Theory of the Physical and Biological World) in which he suggested that the physical-material world is governed by entropy and causality, while the biological world is governed by syntropy and retrocausality.[5]

We cannot see the future and therefore retrocausality is invisible! The dual energy solution suggests the presence of a visible reality (causal and entropic) and an invisible one (retrocausal and syntropic).

---

[5] Fantappiè L., *Principi di una teoria unitaria del mondo fisico e biologico.* Humanitas Nova, Roma 1944.

The first law of thermodynamics states that energy is a unity that cannot be created or destroyed, but only transformed, and the energy-momentum-mass equation shows that this unity has two components: entropy and syntropy. We can therefore write:

$$1 = Entropy + Syntropy \qquad Syntropy = 1 - Entropy$$

where syntropy is the complement of entropy! Life lies between these two components: one visible and the other invisible, one entropic and the other syntropic, and this can be portrayed using a seesaw.

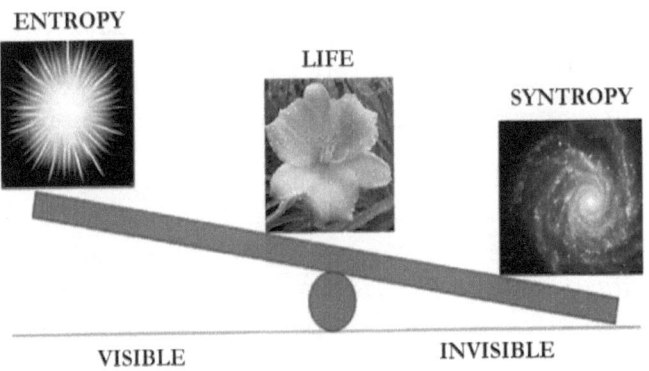

We cannot see the future and therefore syntropy is invisible!

An example is provided by gravity. We continually experience gravity, but we cannot see it. According to the dual time energy solution gravity is a force that diverges backwards in time and, for us moving forward in time, is a converging force. The fact that gravity is invisible is known to all, but that it diverges from the future is known to few.

Can we prove it?

Yes, and it's quite simple. If gravity propagates from the future its speed must exceed that of light. Tom van Flandern (1940-2009), an American astronomer specialized in celestial mechanics, developed a series of procedures to measure the speed of gravity propagation[6,7,8].

[6] Van Flander T. (1996), *Possible New Properties of Gravity*, Astrophysics and Space Science 244:249-261.
[7] Van Flander T. (1998), *The Speed of Gravity What the Experiments Say*, Physics Letters A 250:1-11.
[8] Van Flandern T. and Vigier J.P. (1999), *The Speed of Gravity – Repeal of the Speed Limit*, Foundations of Physics 32:1031-1068.

In the case of light, which has a constant speed of about 300,000 kilometers per second, we observe the phenomenon of aberration. Sunlight takes about 500 seconds to reach the Earth. So when it arrives, we see the Sun in the sky position it occupied 500 seconds before. This difference is equivalent to about 20 seconds of arc, a large amount for astronomers. Sunlight strikes the Earth from a slightly shifted angle and this shift is called aberration.

If the speed of gravity propagation were limited, one would expect to observe aberration in gravity measurements. Gravity should be maximum in the position occupied by the Sun when gravity left the Sun. Instead, observations indicate that there is no detectable delay in the propagation of gravity from the Sun to the Earth. The direction of the gravitational attraction of the Sun is exactly towards the position in which the Sun is, not towards a previous position, and this shows that the speed of propagation of gravity is infinite.

Instant propagation of gravity can only be explained if we accept that gravity is a force that diverges backwards in time, a physical manifestations of syntropy.

Fantappiè failed to prove his theory, since the experimental method requires the manipulation of causes before observing their effects.

Recently, random event generators (REG) have become available. These systems allow to perform experiments in which causes are manipulated after their effects: in the future.

The first experimental study on retrocausality, by Dean Radin of the ION (Institute of Noetic Sciences)[9], measured heart rate, skin conductance and blood pressure in subjects who were presented with blank images for 5 seconds followed by images that, based on a random event generator, could be neutral or emotional. The results showed a significant activation of the parameters of the

---

[9] Radin D.I. (1997), *Unconscious perception of future emotions: An experiment in presentiment*, Journal of Scientific Exploration, 11(2): 163-180.

autonomic nervous system, before the presentation of emotional images.

In 2003, Spottiswoode and May, of the Cognitive Science Laboratory, replicated this experiment by performing a series of controls to study possible artifacts and alternative explanations. The results confirmed those already obtained by Radin[10]. Similar results were obtained by other authors, such as McCraty, Atkinson and Bradley[11], Radin and Schlitz[12] and May, Paulinyi and Vassy[13], always using the parameters of the autonomic nervous system.

---

[10] Spottiswoode P (2003) e May E, *Skin Conductance Prestimulus Response: Analyses, Artifacts and a Pilot Study*, Journal of Scientific Exploration, 2003, 17(4): 617-641.
[11] McCratly R (2004), Atkinson M e Bradely RT, *Electrophysiological Evidence of Intuition: Part 1*, Journal of Alternative and Complementary Medicine; 2004, 10(1): 133-143.
[12] Radin DI (2005) e Schlitz MJ, *Gut feelings, intuition, and emotions: An exploratory study*, Journal of Alternative and Complementary Medicine, 2005, 11(4): 85-91.
[13] May EC (2005), Paulinyi T e Vassy Z, *Anomalous Anticipatory Skin Conductance Response to Acoustic Stimuli: Experimental Results and Speculation about a Mechanism*, The Journal of Alternative and Complementary Medicine. August 2005, 11(4): 695-702.

Daryl Bem, psychologist and professor at the Cornell University, describes nine classic experiments conducted in the retrocausal mode in order to get the effects first rather than after the stimulus. For example, in a priming experiment, the subject is asked to judge whether the image is positive (pleasant) or negative (unpleasant) by pressing a button as quickly as possible. The reaction time is recorded.[14]

Just before the positive or negative image, a word is presented briefly, below the threshold so that it is not perceptible at a conscious level. This word is called *"prime"* and it has been observed that subjects tend to respond more quickly when the prime is congruent with the following image, whether it is a positive or negative image, while the reactions become slower when they are not congruent, for example when the word is positive while the image is negative.

---

[14] Bem D (2011), *Feeling the future: Experimental evidence for anomalous retroactive influences on cognition and affect*, Journal of Personality and Social Psychology, Jan 31, 2011.

In retro-priming experiments, the usual stimulus procedure takes place later, rather than before the subject responds, based on the hypothesis that this "inverse" procedure can retrocausally influence the answers. The experiments were conducted on more than a thousand subjects and showed retrocausal effects with statistical significance of a possibility on 134,000,000,000 of being mistaken when affirming the existence of the retrocausal effect.

Syntropy explains these results in the following way: *"Since life feeds on syntropy, and syntropy flows backward in time, the parameters of the autonomic nervous system that support vital functions must react in advance to future stimuli."*

As part of her doctoral thesis in cognitive psychology, Antonella Vannini conducted four experiments using heart rate measurements to study the retrocausal effect.[15]

---

[15] Vannini A. e Di Corpo U., Retrocausalità, esperimenti e teoria, https://www.amazon.it/dp/1520892527

Each experimental trial was divided into 3 phases:

Phase 1 Presentation of stimuli and measurement of heart rate				Phase 2 Choice	Phase 3 Random selection
Blue	Green	Red	Yellow	Blue/Green/Red/Yellow	Red
					Target
4 seconds HR01 HR02 HR03 HR04	4 seconds HR01 HR02 HR03 HR04	4 seconds HR01 HR02 HR03 HR04	4 seconds HR01 HR02 HR03 HR04		Feedback

— *Phase 1,* in which 4 colors were displayed one after the other on the computer screen. The subject had to look at these colors and during their presentation the heart rate was measured.

— *Phase 2,* in which an image with 4 colored bars was displayed and the subject had to try to guess the color that the computer would have selected.

— *Phase 3,* in which the computer randomly selected the color and showed it full screen.

The hypothesis was that in the case of a retrocausal effect differences should be observed among the heart rates measured in phase 1 in correlation with the target color selected in phase 3 from the computer.

In the absence of the retrocausal effect, the heart rates differences associated with each color of the target stimulus should have varied around the zero value (0).

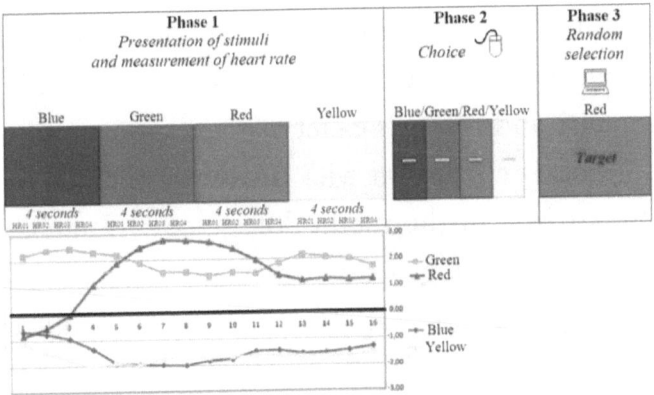

Retrocausal effect observed on a subject

Instead, a marked difference was observed!

In some subjects the heart rate increased when the target color was blue and decreased when the target was green. In others exactly the opposite was observed.

Performing data analysis within each subject, the retrocausal effect was clear. But, when the analysis was conducted in a classical way, adding the effects observed among several subjects, opposite effects canceled.

This suggested that when studying retrocausal effects parametric statistical techniques such as analysis of variance (ANOVA) or Student's t are not suitable, while nonparametric techniques such as Chi Square and Fisher's exact test are appropriate.

This is consistent with the division made by Stuart Mill in the methodology of differences and methodology of concomitant variations.[16]

Mill showed that causality can be studied using:

— The <u>methodology of differences</u>: "*If an element of difference is introduced in two initially similar groups, the differences that are observed can only be attributed to this single element that was introduced.*"

— The <u>methodology of concomitant variations</u>: "*When two phenomena vary*

---

[16] Stuart Mill, *A System of Logic*, 1843.

concomitantly, one may be the cause of the other or both are united by the same cause."

The study of syntropic phenomena requires the use of the methodology of concomitant variations[17] where the information is translated into dichotomous variables (yes/no). This allows to analyze together quantitative and qualitative, objective and subjective information and to manage an unlimited number of variables simultaneously.

---

[17] See: www.amazon.com/dp/1520326637 and www.sintropia.it/sintropia.ds.zip

THERMODYNAMICS
AND
BIODYNAMICS

We are used to the fact that causes always precede their effects. But the energy-momentum-mass equation implies three types of time:

— *Causal time:* when the positive time solution prevails, ie when systems diverge, as is the case of our expanding universe, entropy dominates, causes always precede their effects and time flows forward, from the past to the future. Since entropy rules, retrocausal effects are not possible, such as light waves that propagate backwards in time or radio signals that are received before being transmitted.

— *Retrocausal time:* when the negative time solution prevails, ie when systems converge, as is the case with black holes, retrocausality dominates, effects always precede causes and time flows backwards, from the future to the past. In these systems no forward effects are possible and this is why no light is emitted from black holes.

– *Supercausal time*: when diverging and converging forces are balanced, as is the case of atoms and quantum mechanics, causality and retrocausality coexist and time is unitary.

This time classification recalls the ancient Greek division into: Kronos, Kairos and Aion.

– *Kronos* describes the sequential causal time, which is familiar to us, made of absolute moments that flow from the past to the future.

– *Kairos* describes the retrocausal time. According to Pythagoras, kairos is at the basis of intuitions, of the ability to feel the future and to choose the most advantageous options.

– *Aion* describes the supercausal time, in

which past, present and future coexist. The time of quantum mechanics, of the subatomic world.

This classification suggests that syntropy and entropy coexist at the quantum level, ie in the Aion, and that the properties of life originate at this level.

A question arises:

How does syntropy flow from the quantum level of matter to the macroscopic level of our physical reality, transforming inorganic matter into organic matter?

In 1925 Wolfgang Pauli discovered the hydrogen bond. In water molecules hydrogen atoms are in an intermediate position between the subatomic (quantum) and molecular (macrocosm) levels and provide a bridge that allows syntropy (cohesive forces) to flow from the micro to the macro. Hydrogen bonds increase cohesive forces (syntropy) and make water different from all other liquids. Because of these cohesive forces ten times stronger than the van der Waals forces that hold the other liquids together, water shows abnormal properties. For example, when it expands, it becomes less dense and floats; on the contrary, the other liquids become denser, heavier and sink when they solidify. The uniqueness of water stems from the cohesive properties of syntropy that allow the construction of networks and structures on a large scale.

Hydrogen bonds let syntropy flow from the subatomic level to the macrocosm level, making water essential for life. Ultimately, water is the lifeblood, the essential element for the manifestation and evolution of any biological structure.

Other peculiarities of water are [18]:

– In liquids the solidification process starts from the bottom, as the hot molecules move upwards, while the cold molecules move downwards. The liquid in the lower

---

[18] Ball P., H$_2$O. *A biography of water*, www.amazon.it/dp/0753810921

part is therefore the first that reaches the solidification temperature; for this reason the liquids solidify starting from the bottom. In the case of water, the opposite is true: water solidifies from the top.

— Water has a much higher thermal capacity than other liquids. Water can absorb large amounts of heat, then released slowly. The amount of heat needed to raise the water temperature is far greater than that required for other liquids.

— When cold water is compressed it becomes more fluid. In contrast, in other liquids the viscosity increases with pressure.

— Friction between the surfaces of solids is usually high, while with ice friction is low and ice surfaces are slippery.

— At temperatures close to freezing, ice surfaces stick together when they come into contact. This mechanism allows snow to compact into snowballs, while it is impossible to produce balls of flour, sugar or other solid materials if you do not use water.

– With water the distance between the melting and the boiling temperatures is very high. Water molecules have high cohesive properties that increase the temperature needed to change water from liquid to gas.

Water is not the only molecule with hydrogen bonds. Also ammonia and hydrofluoric acid form hydrogen bonds and these molecules show anomalous properties similar to water. However, water produces a higher number of hydrogen bonds and this determines the high cohesive properties of water that bind the molecules into large and dynamic labyrinths.

Other molecules forming hydrogen bonds fail to construct complex networks and structures in space. Hydrogen bonds impose extremely unusual structural constraints for a liquid. An example of these constraints is provided by snow crystals. However, when water freezes, the mechanism of the hydrogen bond stops and the flow of syntropy from the micro to the macro also stops, bringing life to death.

Hydrogen bonds make water essential for life, water provides syntropy to living systems. If life ever starts on another planet, surely water would be needed. Water is the only means by which life draws syntropy from the quantum level. Consequently, it is the indispensable element for the origin and evolution of any biological structure.

Hydrogen bonds impose structural constraints that are extremely unusual for a liquid, and these in turn affect physical properties such as density, heat capacity and heat conduction, as well as the way water receives within it solute molecules.

When water is super cooled to the experimental limit of -38°C, its thermal capacity increases considerably. At the theoretical limit of -45°C the thermal capacity of water becomes infinite; water could absorb infinite amounts of heat without increasing in temperature. At this theoretical limit, even the slightest increase in pressure would make water disappear, similarly to what happens with black holes in which temporal inversion makes matter disappear.

The syntropic properties of water suggest that water is constantly under the effect of retrocausal forces. This would explain why it is so difficult to predict the behavior of water molecules even in a small glass.

Based on these considerations, in February 2011 with Antonella Vannini I wrote an article for the Journal of Cosmology commenting on an article by dr. Richard Hoover[19] of NASA Marshall Space Flight Center.

Dr. Hoover discovered microfossils, similar to cyanobacteria, in internal sections of comet meteorites and, using electron microscopy and a series of other measures, concluded that they originated from these meteors, ie comets.

According to syntropy, life is a general law of the universe which requires the presence of water to manifest itself. A characteristic of comets is that they are rich in ice which, in the vicinity of the Sun, melts and becomes water; therefore in our article[20] we suggested that,

---

[19] Hoover R (2001), *Fossils of Cyanobacteria in CI1 Carbonaceous Meteorites, Journal of Cosmology*, 2011, http://journalofcosmology.com/Life100.html

according to syntropy, living organisms can originate in extreme conditions, such as those of comets, and that the discovery of Dr. Hoover of cyanobacteria microfossils in meteorites is consistent with the theory of syntropy.

About energy, a law that governs all natural phenomena is energy conservation. This law tells that the amount of energy does not change in the transformations it undergoes. We can calculate the amount of energy and after any processing if we calculate again the amount of energy it is always the same.[21] This is the first law of thermodynamics which states that: *"Energy cannot be created or destroyed, but only transformed."*

Thermodynamics is the branch of physics that studies the behavior of energy, of which heat is a form.

Born from the works of Boyle, Boltzmann,

---

[20] Vannini A (2011) and Di Corpo U, *Extraterrestrial Life, Syntropy and Water*, Journal of Cosmology, http://journalofcosmology.com/Life101.html#18

[21] Feynman R (1965), *The Feynman Lectures on Physics*, California Institute of Technology, 1965, 3.

Clausius and Carnot it identifies three principles, which we here reword according to the law of syntropy:

1. *Principle of energy conservation*: energy can neither be created nor destroyed, but only transformed.
2. *Principle of entropy*, in expanding systems entropy is the quantity of energy that is lost into the environment.
3. *Principle of heat death*, in expanding systems entropy is irreversible, energy dispersion cannot decrease.

Entropy identifies the tendency of physical systems to evolve towards "*heat death*" and the homogeneous distributions and the destruction of all forms of organization. Nevertheless, living systems show the opposite tendency, they evolve towards more complex forms of organization. The iron law of entropy seems contradicted by life. Instead of tending towards homogeneity and disorder, life evolves towards ever more complex forms of organization capable of keeping away from heat death.

The paradox of how life can emerge in a universe governed by the law of entropy, is continually debated by biologists and physicists.

Erwin Schrödinger (Nobel prize for physics), answering the question about what allows life to contrast entropy, replied that life feeds on negative entropy, thus affirming the need for a second type of energy with symmetrical properties to those of physical energy.[22]

It is though important to note that while negentropy is defined with no reference to the direction of time, syntropy is defined as an anticipatory force, complementary to entropy.

$$Entropy(^{+t})=1\text{-}syntropy(^{\text{-}t})$$

This implies a profound shift of paradigm, from the mechanical to the supercausal paradigm.

The same conclusion was reached by

---

[22] Schrödinger E. (1944), *What is life?*
http://whatislife.stanford.edu/LoCo_files/What-is-Life.pdf

Albert Szent-Gyorgyi, (Nobel Prize for physiology in 1937 and discoverer of vitamin C). He borrowed the term syntropy from Fantappiè and he postulated the existence of a complementary force to entropy, a force which causes living things to reach higher and higher levels of organization, order and dynamic harmony:

"It is impossible to explain the qualities of organization and order of living systems starting from the entropic laws of the macrocosm. This is one of the paradoxes of modern biology: the properties of living systems are opposed to the law of entropy that governs the macrocosm ... One of the main differences between the amoebas and humans is the increase in complexity which presupposes the existence of a mechanism that is able to counteract the law of entropy. In other words there must be a force that is able to counteract the universal tendency of matter towards chaos and energy towards heat death. Life continuously shows a decrease in entropy and an increase in its internal complexity and often in the complexity of the environment, in direct opposition to the law of entropy ... We observe a profound difference between the organic and inorganic

systems ... as a man of science I cannot believe that the laws of physics lose their validity as soon as we enter the living systems. The law of entropy does not govern living systems."[23]

The discovery of syntropy requires that we expand thermodynamics to a new set of laws which we here name biodynamics. What we need to add is the following:

4. *Principle of syntropy*, in converging systems energy is absorbed increasing differentiation and complexity. Syntropy is the magnitude with which energy concentration, the increase in differentiation and complexity are measured.

5. *Principle of life* in isolated systems placed in converging systems syntropy is irreversible, energy concentration cannot decrease.

6. *Life as a general law of the universe.* Life manifests whenever the properties of the

---

[23] Szent-Gyorgyi A (1977), *Drive in Living Matter to Perfect Itself*, Synthesis 1977, 1(1): 14-26.

quantum world flow into the macro world thanks to the water molecule.

This last statement is now supported by the fact that the functioning of living systems is largely influenced by quantum events: the length and strength of hydrogen bonds, the transmission of electrical signals in microtubules, the action of DNA, the folding of proteins. Water provides the mean for the syntropic properties of the quantum world to flow into the macro level and change inorganic into organic matter.

The importance of water for life has always been known and it is not a coincidence that living organisms are mainly made of water. The human body consists of 75% water and only 25% of solid matter.

DYNAMIC BALANCE
BETWEEN
ENTROPY AND SYNTROPY

The first law of thermodynamics states that energy is a unity that cannot be created or destroyed, but only transformed. Entropy and syntropy are the two sides of this unity, linked together in a dynamic process of energy transformation. Entropy and syntropy cannot exist without each other. This dynamic interaction pervades all aspects of the universe and that is why everything vibrates and everything is dual.

In 1665, the Dutch mathematician and physicist Christian Huygens, among the first to postulate the wave theory of light, observed that, putting side by side two pendulums, these tended to tune their oscillation as if "*they wanted to take the same rhythm.*" Huygens discovered the phenomenon we now call resonance. In the case of two pendulums, it is said that one makes the other resonate at its own frequency.

All the manifestations of the universe are a continuous vibration between polarities: converging and diverging, syntropy and entropy, absorbers and emitters.

In life, this takes the form of waves, pulsations and rhythms: the pulsations of the heart, the phases of the breath, light and sound waves.

All aspects of reality vibrate and these vibrations create resonances. An example is provided by tuning forks that vibrate at a frequency of 440 Hz. When a vibrating tuning fork is placed near a *"silent"* tuning fork, this second tuning fork begins to vibrate. Tuning forks vibrate only when exposed to a sound with their own resonance.

Resonance is the principle used by radios to tune to a specific station. Tuning to a frequency allows to receive only the information sent with that frequency, all other information is not accessible.

The same happens with life. We only perceive what vibrates at our own frequency. This resonance process allows information to flow. Every person, every event and every situation is associated with a specific vibration. We communicate easily with people who have the same vibration as ours, while communication is more difficult with others.

Individuals who resonate in the same way can easily establish lasting bonds. For example, young people who have had problems with abandonment, violence and abuse in their families tend to attract without knowing each other's history.

Resonance leads people to recognize themselves and to share feelings and information. This empathic communication often takes place at an unconscious level.

We constantly experience resonance. We can talk to more people on the same subject, using the same words, the same gestures and the same emphasis, and with some we feel that communication is full, while with others we feel that communication is empty.

Resonance allows to communicate at a deeper level. When we resonate we feel that communication is intense and profound.

Everything is a vibration between the past and the future. The energy-momentum-mass equation describes the present as the interaction of causes that act from the past (causality) and attractors that act from the future (retrocausality).

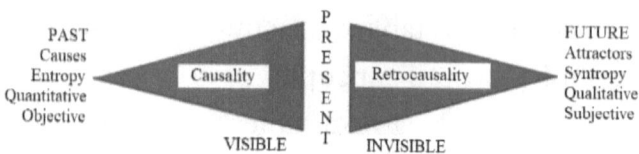

Causes are quantitative and objective and their effects are regulated by the law of entropy. Instead attractors are usually perceived in a qualitative and subjective way. Their effects are governed by the law of syntropy.

On November 24, 1803, Thomas Young demonstrated that light propagates as waves:

"The experiment I'm about to talk can be repeated with great ease, as long as the Sun is shining and with an instrument within everyone's reach."

Young's experiment is very simple. A sunbeam passes through the slit of a screen

(S1), then reaches a second screen (S2) with two slits.

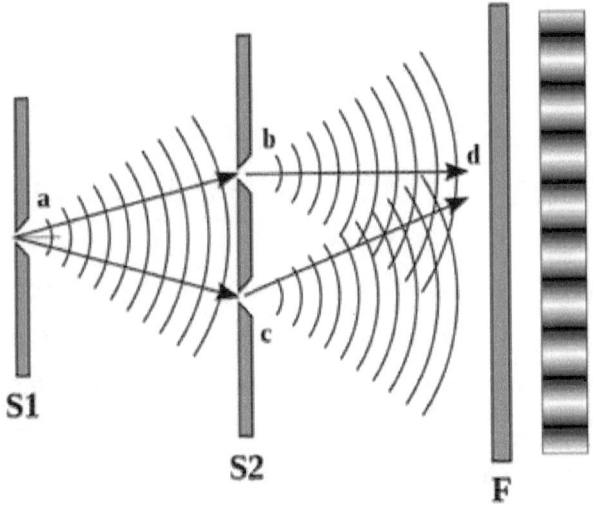

Thomas Young's double-slit experiment

The light that passes through the two slits of the second screen finally ends up on the white screen F, where it creates a figure of lights and shadows. If the light were made of particles, two points of light should be observed at the height of the two slits. Instead we observe a figure in which dark bands and light bands alternate.

Young explained this result as a demonstration of the fact that light propagates through the two slits as waves. These waves give rise to luminous bands at the points where they add up, that is, where there is constructive interference, while they give rise to dark bands where they do not add up, where there is destructive interference.

Everything went well until the end of the nineteenth century, when physicists faced a paradox. Maxwell's equations led to predicting that a black body, an object that absorbs electromagnetic radiations, must emit ultraviolet frequencies with infinite power peaks. Fortunately, this did not happen! This prediction, known as the ultraviolet catastrophe, has never been observed.

The answer was provided by Max Planck on December 14, 1900. In an article that he presented to the German Physics Society, Planck suggested that energy does not propagate in the form of waves, but as multiples of fundamental units, which he called quantum. A quantum can be more or less small depending on the frequency of vibration of the atom. Under the size of the quantum energy does not propagate. This avoids the formation of infinite peaks and solves the paradox of the ultraviolet catastrophe.

In 1905 Einstein explained the behavior of the photoelectric effect considering light made of quanta rather than waves. The photoelectric effect is that when light rays strike a metal, the metal emits electrons. However, up to a certain threshold the metal does not emit electrons and above this threshold it emits electrons whose energy remains constant. The wave theory of light cannot explain this behavior.

Einstein suggested that light, previously considered only as an electromagnetic wave, could be described in terms of quanta, particles we now call photons. The explanation provided by Einstein treats light in terms of particle beams, rather than in terms of waves, and has paved the way for the wave-particle duality.

Today, the exact equivalent of Young's experiment can be conducted using an electron beam. The electrons launched in a double-slit experiment produce an interference pattern on the detector screen and must therefore propagate as waves. However, upon arrival, they generate a point of light, behaving like particles.

If the electrons were particles, they would go through one or the other of the two slits; however the interference shows that they behave like waves that go through the two slits simultaneously.

According to Richard Feynman the central mystery of quantum mechanics is hidden in the double-slit experiment:

"It is a phenomenon in which it is impossible, absolutely impossible, to find a classical explanation, and which represents well the nucleus of quantum mechanics. In reality, it contains the only mystery (...) The fundamental peculiarities of all quantum mechanics."

The wave-particle duality supports the theory of syntropy which states that causality and retrocausality constantly interact and that nothing happens without the contribution of both. The past manifests itself as particles (causality), while the future as waves (retrocausality). An emitter with particle properties and an absorber with wave properties are required for light to propagate.

Quantum mechanics tries to explain this duality by keeping the manifestations of waves and particles separate. For example, the Copenhagen interpretation says that the particle turns into a wave and then the wave collapses back into a particle. According to syntropy the dual nature wave-particle coexists and it is inseparable, since all the manifestations of the universe are the result of the interaction between entropy and syntropy, between past and future, between emitters and absorbers.

- Diverging and converging cycles

The dynamic balance between entropy and syntropy presumes that any system vibrates between peaks of expansion and contraction:

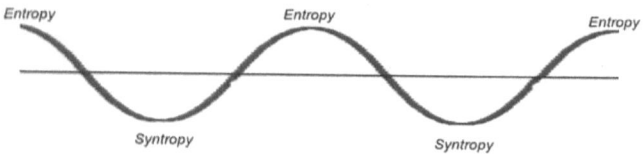

These cycles can be observed in any system and at any level, from the quantum level to the macro level and at the cosmological level where it supports Einstein's cosmological model of infinite cycles of Big Bang and Big Crunch.

The first formulation of the Big Bang theory dates back to 1927 but was generally accepted only in 1964 when many scientists were convinced that observations confirmed that an event such as the Big Bang took place. Georges Lemaître, a Belgian Catholic priest and physicist, developed the Big Bang equations and suggested that the increase in the distance of galaxies was due to the expansion of the universe.

He discovered a proportionality between distance and spectral displacement (now known as the Hubble law).

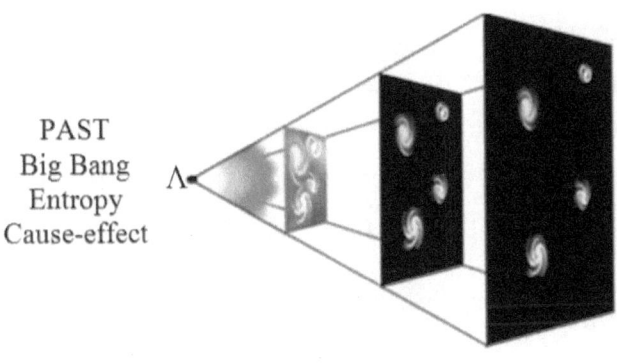

PAST
Big Bang
Entropy
Cause-effect

In 1929 Edwin Hubble and Milton Humason noted that the distance of galaxies is proportional to their redshift, the shift towards the lower frequencies of the light spectrum. This usually happens when the light source moves away from the observer or when the observer moves away from the source. The spectrum of the light emitted by far away galaxies, quasars or supernovas, appears shifted to lower frequencies. Since red is the lowest frequency of the visible light, the phenomenon has received the name of redshift, even if it is used in connection with any frequency, including radio frequencies.

The redshift phenomenon indicates that galaxies are moving away from each other and, more generally, that the universe is in an expansion phase. Furthermore, red-shift measurements show that galaxies and star clusters move away from a common point in space and that the farther they are from this point, the greater their speed.

Since the distance between the galaxy clusters is increasing, it is possible to deduce, going backwards in time, density and increasingly higher temperatures until reaching a point where values tend to infinity and the physical laws of positive time energy are no longer valid.

In cosmology, the Big Crunch is a hypothesis about the fate of the universe. This hypothesis is symmetrical to the Big Bang and claims that the universe will stop expanding and will begin to collapse on itself. Gravitational forces will prevent the universe from expanding to infinity and the universe will converge.

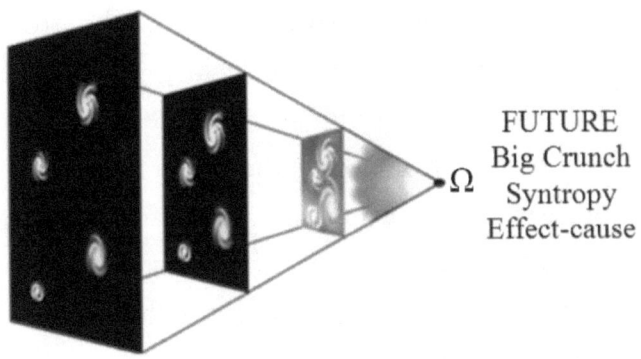

FUTURE
Big Crunch
Syntropy
Effect-cause

Ω

The contraction will appear very different from the expansion. While the early universe was highly uniform, a shrinking universe will always be more diverse and complex. Eventually all the matter will collapse into black holes, which will then unite, creating a unified black hole, the singularity of the Big Crunch. The Big Crunch theory proposes that the universe can collapse in the state it started and then start another Big Bang. In this way the universe would last forever, going through an infinite sequence of expansion cycles (Big Bang) and contraction cycles (Big Crunch).

Recent observations, particularly that of distant supernovae, led to the idea that the expansion of the universe is not slowed down by gravity, but rather is accelerating.

In 1998, the measurement of light from distant stars led to the conclusion that the universe is expanding at an increasing rate. The observation of the red shift of supernovae suggests that they are moving away more quickly as the universe ages. According to these observations, the universe seems to expand at an increasing rate. This contradicts the Big Crunch hypothesis.

In an attempt to explain these observations, physicists have introduced the idea of dark energy, of a dark fluid or phantom energy. The most important property of dark energy would be to exert a relatively homogeneously distributed negative pressure in space, a kind of anti-gravitational force that is moving galaxies away. This mysterious anti-gravitational force is considered a cosmological constant, which will lead the universe to expand exponentially. However, until today no one knows what dark energy is or where it comes from.

Conversely, Syntropy suggests that the observed increase in the rate of expansion of the universe is not due to dark energy or other mysterious anti-gravitational forces, but to the fact that time is slowing down.

In June 2012, José Senovilla, Marc Mars and Raül Vera from the University of Bilbao and the University of Salamanca published an article in the journal Physical Review D in which they dismissed dark energy as an invention. Senovilla says that acceleration is a blunder caused by time that gradually slows down:

"We do not say that the expansion of the universe is an illusion, what we say is that the acceleration of this expansion is an illusion. [...] in our equations we have naively maintained the flow of time constant, so the simple models we have built show that an acceleration of the expansion occurs."

The corollary of Senovilla's group is that dark energy does not exist and that we have been deceived into thinking that the expansion of the universe is accelerating, when instead it is time that is slowing down.

On a daily basis, this change is not perceptible, but when measurements are based on light emitted by stars exploded billions of years ago it is easily detectable.

Astronomers measure the rate of expansion of the universe using the red-shift technique and stars that move farther away appear to have a more marked red color. However, they treat time as a constant.

But if time slows down it becomes a spatial dimension. So the most distant and ancient stars would seem to accelerate. Professor Senovilla says:

"Our calculations show that we would fall into the illusion of thinking that the expansion of the universe is accelerating."

Although radical and in many ways unprecedented, this interpretation is not without its supporters. Gary Gibbons, a cosmologist at the University of Cambridge, says:

"We believe that time has emerged during the Big Bang, and if time can emerge, it can also disappear - this is just the opposite effect."

The dual time solution of the energy-momentum-mass equation suggests a cosmological interpretation of the universe that vibrates between peaks of expansion and contraction. The fastest is the expansion and the fastest is the forward flow of time, the fastest is the contraction and the fastest is the flow of time backwards.

The Big Bang is governed by positive time and entropy, that is energy and matter that diverge from an initial point, while the Big Crunch is governed by negative time and syntropy, that is energy and matter that converge towards a point of final density and infinite temperature.

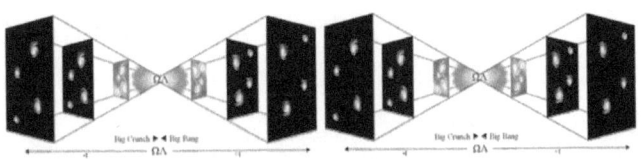

Big Bang and Big Crunch cycles

The Big Bang is indicated with the first letter Λ = Alpha (the beginning), of the Greek alphabet, while the Big Crunch with the letter

Ω = Omega (the end).

The question that is often heard among cosmologists is why we live in a universe predominantly made of matter. What happened to antimatter? This question is easily answered when we consider the dual time solution. At the time of the Big Bang the amount of matter and antimatter was the same, but antimatter moved backward in time, while matter moved forward in time, thus preventing their annihilation.

According to this interpretation, the universe is made of an equal amount of matter and antimatter, which move in opposite time directions. Two symmetrical planes that influence each other in the continuous interaction between diverging and converging forces, causality and retrocausality, entropy and syntropy, heat and gravity, particles and waves.

All that diverges is governed by the positive time solution, while all that converges is governed by the negative time solution. The physical and material plane continuously interacts with the non-physical and intangible plane of antimatter that propagates backwards in time.

The complexity of the physical universe is a consequence of the interaction between matter and energy with the cohesive forces of anti-matter and anti-energy.

The same model can be applied to atoms, small universes that expand and contract at immense speeds, where each vibration corresponds to an entire Big-Bang/Big-Crunch cycle. During the expansion phase the atom can emit an energy packet (a quantum), while during the contraction phase it can absorb an energy packet. Our universe would therefore be a Boolean universe made of packets, like computer bits.

In the same way our universe could be considered an atom of a much larger universe, and this in turn an atom of an even larger universe and so on towards the infinitely large and towards the infinitely small.

- Weather and temperature cycles

There is no doubt that CO_2, temperatures and sea levels are increasing. But if we look at it from a broader perspective, the picture seems affected by cycles. In this regard, the past can tell us a lot about the future.

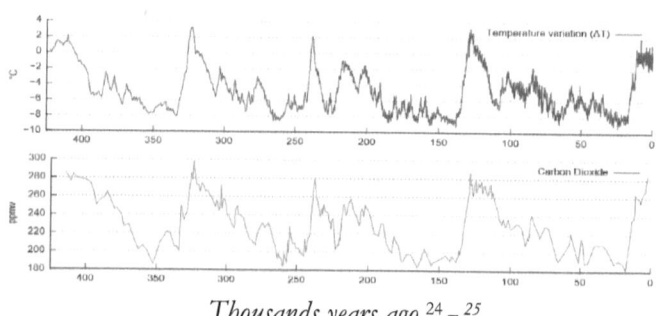

Thousands years ago [24] – [25]

When we examine data on carbon dioxide (CO_2) and temperatures, that are available for the last 800 thousand years, we see that the Earth goes through regular cycles of warm

---

[24] Wikipedia:
https://en.wikipedia.org/wiki/Ice_age#/media/File:Vostok_Petit_data.svg
[25] CDIAC – Carbon Dioxide Information Analysis Center
http://cdiac.ornl.gov/images/air_bubbles_historical.jpg
http://cdiac.ornl.gov/trends/co2/ice_core_co2.html

periods, associated with increasing levels of CO_2, and ice ages of about 100 thousand years. The warm interglacial periods (with average temperatures above 0°C) last about 10 thousand years.

CO_2 is produced by life activities such as breathing and decomposition, industrial activities and the use of fossil fuels such as coal, oil and natural gas. CO_2 levels similar to or higher than the present one indicate that in addition to natural sources, industrial activities existed. CO_2 traps heat providing a "warm blanket" to the planet. However, this "greenhouse effect" was never sufficient to compensate for the lowering temperatures of the ice age.

Civilizations that preceded us in previous interglacial periods seem to have used CO_2 to counteract the reduction in temperatures of the ice age. But none were successful.

The scenario is quite simple! When the ice age begins, temperatures fall by an average of 10/12 degrees. This drop in temperatures is slowed by high CO_2 levels. But when civilizations succumb to the ice, CO_2 levels decrease and polar ice caps expand to reach 3 kilometers at latitudes like Rome and New York. Oceans levels drop by about 300 meters and civilizations are forced to migrate towards the equatorial strip and occupy the land that was previously covered by the oceans.

At the end of the ice age the increase in temperatures is sudden. This causes the polar ice caps to melt into huge interglacial lakes. The banks of these lakes suddenly break, bringing water to increase the levels of the oceans of tens of meters at a time, wiping out what was left of the previous civilizations. Reports of these floods can be found in all the traditions and date back to around 12,000 years ago.

The warm period in which we live began 12,000 years ago and now we are at the end, we are about to enter the next ice age!

Why are glacial cycles so regular?

Because the Sun is not constant in its emissions.

The solar cycles were discovered in 1843 by

Samuel Heinrich Schwabe who after 17 years of observations noted a periodic change in the average number of sunspots in a progression that follows an 11-year cycle. Scientists were baffled by the fact that each cycle was a bit different and no model could explain these fluctuations.

In 2015 it was discovered that these fluctuations are caused by a double dynamo effect between two layers of the Sun, one near the surface and one inside its convection area. This model explains the irregularities of the past and predicts what will happen in the future.

Valentina Zharkova, one of the discoverers of this model, describes the results in this way:

"We found magnetic waves that appear in pairs, originating from two different layers within the Sun. Both have a cycle of about 11 years, even if they are slightly out of phase. During the cycle, the waves float between the northern and southern hemispheres of the Sun. Combining these waves and comparing them with the real data for the past solar cycles, we found that our predictions are 97% accurate."[26]

Using this model to predict the future we see that the pairs of waves will become increasingly out of phase during cycle 25, which reaches its peak in 2022. In cycle 26, which covers the decade from 2030 to 2040, the two waves will become totally out of phase and this will cause a significant reduction in solar emissions.

"In cycle 26, the two waves are opposed to each other, with their peak at the same time but in opposite hemispheres of the Sun. Their interference will be destructive and will cancel each other out ... when the waves are in phase, they can show a strong resonance, and we have strong solar activity. When they are out of phase, we have solar minima."

The Sun is falling asleep and this is evident in the data available on the space weather website: www.spaceweatherlive.com

---

[26] Royal Astronomical Society – *Irregular heartbeat of the Sun driven by double dynamo*
https://www.ras.org.uk/news-and-press/2680-irregular-heartbeat-of-the-sun-driven-by-double-dynamo

Solar Cycle progression - Sunspot number

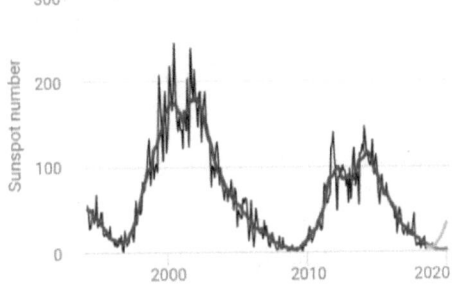

- ● Monthly mean Sunspot Number
- → Monthly Smoothed Sunspot Number
- ■ Predicted Sunspot Number (Standard curves m...
- ☀ Predicted Sunspot Number (Combined method)

The last drop of 1.3 degrees Celsius in global temperatures led to the mini-glaciation of 1645-1715, a period known as the Maunder minimum, in which the hot seasons were short and there was a lack of food.

Zharkova expects a 60% drop in solar activity in the 2030-2040 period.

When solar emissions decrease, the magnetic shield that protects the Earth weakens and cosmic rays enter the core, activating magma and causing strong earthquakes and volcanic eruptions. More than a million volcanoes lie under the sea level against 15,000 on land. Increased eruptions of submarine volcanoes rise ocean temperatures, causing extreme weather conditions such as violent hurricanes and the increase in the amount of water vapor in the atmosphere.

High levels of CO_2 associated with warm interglacial periods suggest the existence of ancient intelligent and industrialized civilizations prior to the last ice age.

Are there traces of these civilizations?

Many archaeological discoveries cannot be explained and remain an enigma for experts. These findings are called *out of place artifacts* (OOPARTS). Artifacts that defy conventional chronology being too advanced for the level of civilization existing at the time, or because they show an intelligent presence before human beings.

In the book "*The Ancient Giants Who Ruled America: The Missing Skeletons and the Great Smithsonian Cover-Up*"[27] Richard Dewhurst

---

[27] Dewhurst R.J., *The Ancient giants Who Ruled America: The Missing Skeletons and the Great – Smithsonian Cover-Up*

presents evidence of an ancient race of giants in North America and the concealment by the Smithsonian Institution.

Thousands of skeletons of giants have been found, particularly in the Mississippi Valley and also ruins of their cities. The book includes more than 100 photographs and illustrations and shows that the Smithsonian Institution came, took the skeletons for further study, and then made them disappear.

In some cases, other government institutions were involved. But the result was always the same: skeletons were removed and disappeared forever.

Why?

OOPARTS and pre-glacial civilizations contradict the narrative that we are the first civilization on this planet.

- Metabolism cycles

Since the concentration of energy cannot take place infinitely, when the limit is reached the process reverses and entropy takes over releasing energy and matter. In turn, the release of energy cannot be infinite, when the limit is reached the process reverses and syntropy takes over concentrating energy and matter.

This process activates an exchange of energy and matter with the environment: syntropy absorbs and organizes, entropy releases and destroys.

Exchange is essential in all living forms, from biological to economic ones.

This continuous exchange is evident in metabolism.

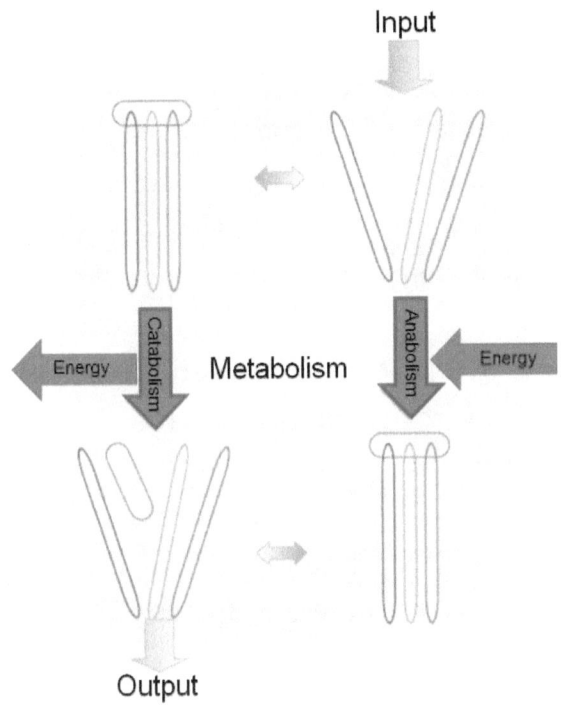

In the form of:

- *anabolism* (ie syntropy) which absorbs energy and leads to the formation of complex biomolecules from simpler ones and nutrients;

- *catabolism* (ie entropy) that decomposes complex biomolecules in structurally simpler ones releasing energy in chemical (ATP) or thermal form.

- In philosophies and religious traditions

The idea of a dynamic balance between two complementary forces, one diverging and one converging, one visible and one invisible, one destructive and one constructive, can be found in many philosophies and religious traditions.

In the Taoist philosophy, for example, all aspects of the universe are regarded as the interplay of two fundamental and complementary principles: *yang*, which is converging, and *yin*, which is diverging.

This is beautifully represented in the Taijitu symbol, which shows the union and interaction of these two principles whose combined action is believed to move all aspects of the universe.

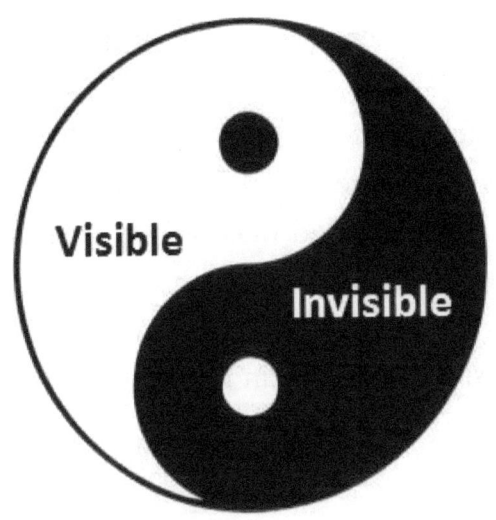

Taijitu symbol

In Hinduism the same law of complementarity is described with the cosmic dance of Shiva and Shakti, where Shakti is the personification of the feminine principle and is the energy of the visible physical world, and Shiva is the masculine principle, the ordering principle or consciousness that transcends the visible world.

As in the Chinese Yin and Yang, each contains an aspect of the other. Shiva would thus represent the organizing properties of syntropy and come from the future, whereas Shakti would represent the disordering properties of entropy and flow from the past. Together they represent the dynamic organizing forces and the primordial cosmic energy that are expressed throughout the entire universe, and one cannot exist without the other. Sometimes they are represented by a single figure called *Ardbanarisvara*, whose right side is male and whose left side is female.

TIME

Let's see how the concept of time has changed from Galilean relativity to Einstein's special relativity.

In 1623 Galileo formulated the law of composition of velocities which is also known as Galilean relativity.

This law stems from the fact that, when inside a system, it is not possible to detect if it is moving with uniform motion. Galileo used the example of a ship travelling at a constant speed, without rocking, on a smooth sea. Any observer below the deck would not be able to tell whether the ship is moving or stationary.

Galileo formulated this concept in his Second Day of the *Dialogue Concerning the Two Chief World Systems*:[28]

---

[28] Galileo Galilei, *Giornata Seconda del suo Dialogo sui Massimi Sistemi del Mondo* (1623)

"*Shut yourself up with some friends in the main cabin below decks on some large ship, and have with you there some flies, butterflies, and other small flying animals. Have a large bowl of water with some fish in it; hang up a bottle that empties drop by drop into a wide vessel beneath it. With the ship standing still, observe carefully how the little animals fly with equal speed to all sides of the cabin. The fish swim indifferently in all directions; the drops fall into the vessel beneath; and, in throwing something to your friend, you need throw it no more strongly in one direction than another, the distances being equal; jumping with your feet together, you pass equal spaces in every direction. When you have observed all these things carefully (though doubtless when the ship is standing still everything must happen in this way), have the ship proceed with any speed you like, so long as the motion is uniform and not fluctuating this way and that. You will discover not the least change in all the effects named, nor could you tell from any of them whether the ship was moving or standing still. In jumping, you will pass on the floor the same spaces as before, nor will you make larger jumps toward the stern than toward the prow even though the ship is moving quite rapidly, despite the fact that during the* time that you are in the air the floor under you will be going in a direction opposite to your jump. In throwing something to your companion, you will need no more force to get it to him whether he is in the direction of the bow or the stern, with yourself situated opposite. The droplets will fall as before into the vessel beneath without dropping toward the stern, although while the drops are in the air the ship runs many spans. The fish in their water will swim toward the front of their bowl with no more effort than toward the back, and will go with equal ease to bait placed anywhere around the edges of the bowl. Finally the butterflies and flies will continue their flights indifferently toward every side, nor will it ever happen that they are concentrated toward the stern, as if tired out from keeping up with the course of the ship, from which they will have been separated during long intervals by keeping themselves in the air. And if smoke is made by burning some incense, it will be seen going up in the form of a little cloud, remaining still and moving no more toward one side than the other. The cause of all these correspondences of effects is the fact that the ship's motion is common to all the things contained in it, and to the air also. That is why I said you should be below decks; for if this took place above in the open air, *which would not follow the course of the ship, more or less noticeable differences would be seen in some of the effects noted."*

Whereas for an observer in the ship it is impossible to conclude whether it is moving or stationary, for an observer on another "inertial system", for example on the seashore and looking to the ship in motion, the speeds of bodies on the ship will add up to the speed of the ship. The Galilean law of composition of velocities consists of a set of rules which are based on the assumption that time is constant and speeds are variables which can add up. For example, if a ship is moving at 20 km/h:

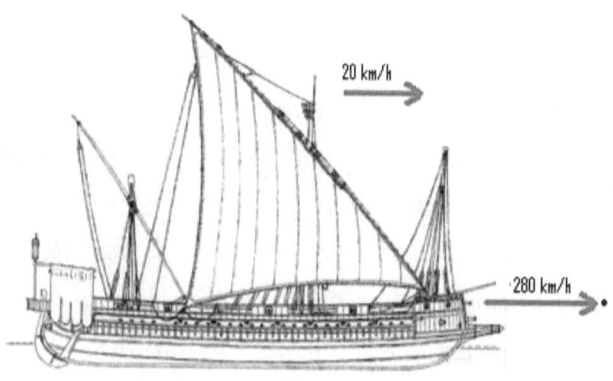

and a cannon ball is fired at 280 km/h in the same direction to the movement of the ship, the observer on the seashore will see the cannon ball move at 300 km/h: 280 km/h of the speed of the cannon ball plus 20 km/h of the speed of the boat.

If the cannon ball were fired in the opposite direction to the movement of the ship the resulting speed would be 260 km/h, 280 km/h of the speed of the cannon ball minus 20 km/h of the speed of the boat (speeds are subtracted because they move in opposite directions).

On the contrary for a sailor on the ship sharing the same movement of the ship, the cannon ball would always move at 280 km/h in any direction he would fire it.

Therefore, if an observer on the seashore sees the cannon ball moving at 300 km/h and the boat in the same direction at 20 km/h he can conclude that the cannon ball was fired at 280 km/h.

Galileo's relativity is based on the principle that when changing inertial system, speeds are added or subtracted. Galileo's relativity allowed to generalize the laws of mechanics.

Two centuries later, in 1881, Albert Michelson began a series of experiments in order to measure the speed of the ether.

The wave theory of light postulated the existence of a substance for the propagation of light waves. It was thought, in fact, that light propagates in an element that permeates the entire universe.

But, the nature of this substance was the source of numerous problems. One was the fact that light required a solid ether and that the very high speed of propagation of light was possible if ether was highly rigid. Then the aberration of the light of the stars indicated that ether had to remain motionless, even at astronomical distances. However, no resistance to the motion of bodies could be attributed to ether.

Earth and the solar system orbit around the center of the galaxy at a speed of 217 km/s. A wind of ether with that speed would therefore have to invest the Earth in the opposite direction to its motion: a variable ether wind according to the latitude, with a peak of 460 m/s at the equator. It was also known the motion of Earth around the Sun at a speed of about 30 km/s.

Albert Michelson devised a tool that enabled to split light into two beams traveling along perpendicular paths which were then made to converge on a screen, where they formed an interference pattern. An ether wind would have resulted in a different speed of light in various directions and, consequently, a sliding movement of the interference fringes when rotating the apparatus with respect to the ether wind direction.

Using this device, now known as the interferometer, in 1881 Michelson accomplished several experiments, but he never detected the minimum displacement in the interference fringes. He published the data and results in the same year.

Michelson interferometer was not sufficiently precise to exclude the existence of the ether and for this reason he asked the cooperation of Edward Morley, who made available his basement for new experiments with an interferometer mounted on a square stone slab of 15 cm and about 5 cm thick floating on liquid mercury, a technique which allowed to maintain horizontal the interferometer device and turn it around a central pin, eliminating any vibration.

A system of mirrors sent the beam of light which followed a path of eight round trips in order to make the beam of light travel as long as possible.

However, even in this new set of experiments there was no trace of ether and the speed of light appeared to be independent of the direction of the path and a little lower to 300,000 km/s. The results were then confirmed by repeating the experiment at a distance of time and place and led to the famous conclusion that the speed of light is constant and that ether does not exist.

The fact that the speed of light is constant undermines Galileo's relativity since the speed of light does not add up to the body that emits it and opened the door to a disturbing scenario, namely that the laws of physics are local and cannot be generalized.

In 1905, analyzing Michelson, Morley and Lorentz' results, Einstein overturned Galileo's relativity according to which time is absolute and speeds are relative.

In order to describe the fact that the speed of light is constant, it was necessary to accept that time is relative. Einstein developed this intuition in his Special Relativity.

Let us imagine, after 500 years, an astronaut on a very fast space ship heading towards Earth at 20,000 km/s who shoots a laser light ray towards Earth (at 300,000 km/s). An observer on Earth will not see the laser light arrive at 320,000 km/s, as Galileo's relativity predicts, but at 300,000 km/s (because the speed of light is constant and does not add up). According to Galileo's relativity, the observer on Earth would expect that the astronaut on the space ship would see the light ray move at 280,000 km/s (300,000 km/s of the speed of light minus 20,000 km/s of the space ship) but, on the contrary, he also sees the laser ray move at 300,000 km/s.

Einstein suggested that what varies is time: when we move in the direction of light our time slows, and for us light continues to move at the same speed.

This leads to the conclusion that approaching the speed of light time would slow down and stop, and if we could move at speeds higher than the speed of light, time would reverse.

In other words, events which happen in the direction in which we are moving become faster, because time slows down, but events which happen in the direction from which we are coming become slower, because time becomes faster.

Einstein came to the conclusion that with light what varies is not the speed, but the flow of time.

Returning to the example of the space ship, when we move in the direction of the light beam, our time slows, and for us the light continues to travel at 300,000 km/s.

In other words, events that happen in the direction in which we move become faster, because time slows down, but events that happen in the direction from which we are moving from become slower because our time speeds up.

In order to explain this situation, Einstein liked to use the example of a lightning which strikes a railway simultaneously in two different points, A and B, far away from each other.

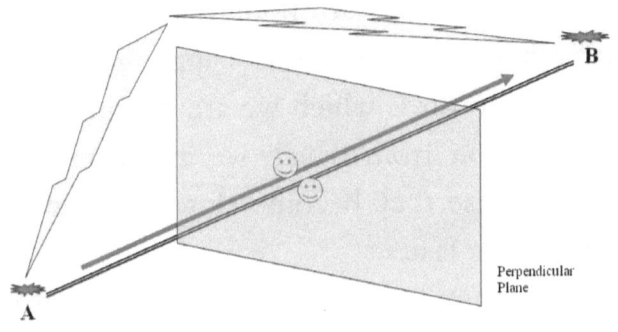

Perpendicular
Plane

A

B

An observer sitting on a bench half-way would see the lightning strike the two points simultaneously, but a second observer on a very fast train moving from A to B passing next to the first observer at the moment in which the lightning strikes the two points would have already experienced the lightning striking point B, but would have not experienced the lightning striking point A.

Even if the two observers share the same point of space at the same moment, they cannot agree on the events which are happening in the direction in which the second observer is moving. Agreeing on the existence of contemporary events is therefore linked to the speed at which the observers are moving.[29]

It is important to note that time flows differently if the event is happening in the direction towards which we are moving, or in the direction from which we are coming: in the first case they become slower and in the second case faster.

This example is limited to two observers; but what happens when we compare more than two observers moving in different directions at high speeds?

The first couple (one on the bench and the other in the train) can reach an agreement only on the contemporary existence of events which happen on a plane perpendicular to the movement of the train.

If we add a third observer moving in another direction, but sharing the same place

[29] Einstein A. (1916), Relativity: The Special and the General Theory. www.amazon.it/dp/048641714X

and moment with the other two observers, they would agree only on the contemporary existence of events placed on a line which unites the two perpendicular planes.

If we add a fourth observer, they would agree only on a point which unites the three perpendicular planes.

If we add a fifth observer, who is not even sharing the same point in space, no agreement would be possible at all.

If we consider that only what happens in the same moment exists (Newton's time concept), we would be forced to conclude that reality does not exist.

In order to re-establish an agreement between the different observers, and in this way the existence of reality, we need to accept the coexistence of events which could be future or past for us, but contemporary for another observer. Extending these considerations, we arrive at the necessary consequence that past, present and future coexist.

Einstein himself found it difficult to accept this consequence of special relativity.

THE COMPASS OF THE HEART

The autonomic nervous system automatically and unconsciously regulates the vital functions of the body, without the need for any voluntary control.

Almost all the visceral functions are under the control of the autonomic nervous system which is divided into the sympathetic and parasympathetic systems. The nerve fibers of these systems do not directly reach the organs, but stop first and form synapses with other neurons in structures called ganglia, from which other nerve fibers form systems, called plexuses, which reach the organs. The sympathetic part of the system is close to the spinal ganglia and forms synapses together with longitudinal fibers, in a tree called the paravertebral chain. The parasympathetic system forms synapses away from the spine and closer to the organs it controls. The ganglia of the sympathetic system are distributed as follows: 3 pairs of intracranial ganglia, located along the trigeminal, 3 pairs of cervical ganglia connected to the heart; 12 pairs of dorsal ganglia connected to the lungs and the solar plexus, 4 pairs of lumbar ganglia that are connected through the solar plexus to the stomach, small intestine, liver, pancreas and kidneys, 4 pairs of ganglia in connection with the rectum, bladder and genital organs.

For a long time it was believed that there was no relationship between the brain and the sympathetic system, but today we know that this relationship exists, is strong and that the brain can act directly on the organs through the mediation of the solar plexus. There is therefore a link between mental states and physical states. For example, sadness acts on the solar plexus through the sympathetic system, generating a vasoconstriction due to the contraction of the arterial system. This contraction caused by sadness hinders blood circulation, thus also affecting digestion and respiration.

People commonly refer to the heart and not to the solar plexus. However, from a physiological point of view, the organ that allows us to perceive our inner feelings is the solar plexus.

Syntropy nourishes the vital functions and is a converging energy that propagates from the future, consequently when the inflow of syntropy is good we feel warmth (ie energy concentration) and well-being in the thoracic area of the autonomic nervous system.

On the contrary, when the inflow is insufficient we feel emptiness, pain and anxiety.

These sensations work like the needle of a compass which points towards the source of syntropy (ie life energy).

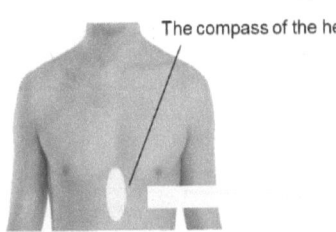
The compass of the heart

The *Attractor*

Unfortunately most people are unaware of how the compass of the heart works and their main concern is to avoid suffering and the unbearable feeling of anxiety. This explains, for example, the mechanism of drug addiction. Substances that act on the autonomic nervous system, such as alcohol and heroin, causing feelings of warmth and wellbeing similar to those that we experience when there is a good inflow of syntropy, can soon become vital.

The compass of the heart points to the source of syntropy, but drugs, alcohol and whatever we use to sedate our suffering reduces our possibility to use the compass of the heart and chose what is beneficial for life.

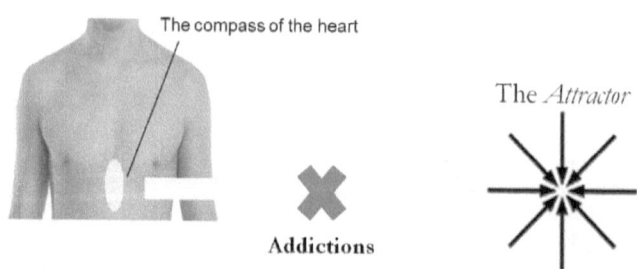

The compass of the heart

Addictions

The *Attractor*

In order to improve the flow of syntropy and promote wellbeing it is therefore essential to abandon any kind of addiction.

While the brain is made of gray matter outside and white matter inside, exactly the opposite is observed in the solar plexus. The gray matter is made up of nerve cells that allow us to think, the white matter is made of nerve fibers, cell extensions, which allow us to feel.

The solar plexus and the brain are the opposite of each other and represent two polarities: the emitter pole and the absorber pole. The same duality that is found between entropy and syntropy.

The solar plexus and the brain are closely connected and from a phylogenetic perspective the brain has developed from the solar plexus. Between the brain and the solar plexus there is a specialization of functions that are completely different and that can only occur when these two polarities are integrated and work in harmony, producing results that are quite extraordinary.

Experiments show that syntropy acts mainly on the solar plexus and is perceived as warmth and well-being. On the contrary, the lack of syntropy is perceived as emptiness and suffering.

Since syntropy propagates backwards in time, feelings of warmth and emptiness help us feel the future and orient our choices towards advantageous goals. The following examples provide some insights into the implications of this backward in time flow:

– The article *"In Battle, Hunches Prove to be Valuable"*, published on the front page of the New York Times on July 28, 2009, describes how experiences associated with intuitions and premonitions helped soldiers save themselves: *"My body suddenly became cold; you know, that feeling of danger, and I started screaming no-no!"* According to syntropy, the attack happens, the soldier experiences fear and death and these feelings of distress propagate backward in time. The soldier in the past feels these as premonitions and is driven to take a different decision, thus avoiding the attack and death. According to the New York Times article, these premonitions have saved more lives than the billions of dollars spent on intelligence.

– William Cox, conducted studies on the number of tickets sold in the United States for commuter trains between 1950 and 1955 and found that in the 28 cases where commuter trains had accidents, fewer tickets were sold[30]. Data analysis was repeated verifying all possible intervening variables, such as bad weather conditions, departure times, day of the week, etc. But no intervening variable was able to explain the correlation between reduced ticket sales and accidents. The reduction of passengers on trains that have accidents is strong, not only from a statistical point of view, but also from a quantitative point of view. According to syntropy, Cox's discoveries can be explained in this way: when people are involved in accidents, the feelings of pain and fear propagate backward in time and can be felt in the past in the form of presentiments and premonitions, which can lead to the

---

[30] Cox WE (1956), *Precognition: An analysis*. Journal of the American Society for Psychical Research, 1956(50): 99-109.

decision not to travel. This propagation of feelings can therefore change the past. In other words, a negative event occurs in the future and informs us in the past, through our inner feelings. Listening to these feelings can help us decide differently and avoid pain and suffering in our future. If we listen to the inner voice, the future can change for the better.

- Among many possible examples: on May 22, 2010 an Air India Express Boeing 737-800 flying between Dubai and Mangalore crashed during landing, killing 158 passengers, only eight survived the accident. Nine passengers, after check-in, felt sick and could not get on board.

In this regard, the neurologist Antonio Damasio, who has studied people affected by decision-making deficits, discovered that feelings contribute to the decision-making process and make advantageous choices possible without having to make advantageous evaluations.[31]

---

[31] Damasio AR (1994), *Descarte's Error. Emotion, Reason,*

Damasio observed that cognitive processes were added to emotional ones, maintaining the centrality of emotions in the decision-making process. This is evident in times of danger: when choices have to be made quickly reason is bypassed.

People with decision making deficit show knowledge but not feelings. Their cognitive functions are intact, but not the emotional ones. They have normal intellect, but are unable to make appropriate decisions. A dissociation between rationality and decision-making skills is observed. The alteration of feelings causes a myopia towards the future. This may be due to neurological lesions or to the use of substances, such as alcohol and heroin, which reduce the perception of our inner feelings.

Feelings of warmth point to the path that leads to well-being and to what is beneficial to life. It is therefore good to choose according to these inner feelings.

When we converge towards the attractor feelings of warmth inform that we are on the

and the Human Brain, Putnam Publishing, 1994.

right path, on the contrary when we diverge we feel void and anxiety.

Intuitions arise from the ability to feel the future and are based on inner feelings not contaminated by drugs, alcohol, habits and fears.

Henri Poincaré, one of the most creative mathematicians of the last century, observed that when faced with a new problem whose solutions can be countless, a rational approach is initially used, but being unable to arrive at the result another type of process is activated.

This process selects the correct solution among the endless possibilities, without the help of rationality.

Poincaré called it intuition (combining the Latin words *in*=inside + *tueri*=glance), and was struck by the fact that they are always accompanied by experiences of truth, beauty, warmth and well-being in the thoracic area:[32]

"Among the large number of possible combinations,

---

[32] Henri Poincaré, Mathematical Creation, from Science et méthode, 1908.

almost all are without interest or utility.
Only those that lead to solving the problem
are illuminated by an interior experience of truth
and beauty."

For Poincaré, intuitions require attention and sensitivity to these inner feelings of truth and beauty, which connect us to the future, to the intelligence of syntropy.

Robert Rosen (1934-1998), theoretical biologist and professor of biophysics at the Dalhousie University, in his book *Anticipatory Systems*[33] wrote:

"I was amazed by the number of anticipatory behaviors observed at all levels of the organization of living systems (...) that behave like real anticipatory systems, systems in which the present state changes according to future states, violating the law of causality according to which changes depend exclusively on past or present causes. We try to explain these behaviors with theories and models that exclude any possibility of

---

[33] Rosen R (1985) *Anticipatory Systems*, Pergamon Press, USA 1985.

anticipation. Without exception, all biological theories and models are classic in the sense that they seek only causes in the past or present."

To make anticipatory behaviors consistent with the idea that causes must always precede effects, predictive models and learning processes are taken into account. But anticipatory behaviors are found also in the simpler forms of life, such as cells, without neural systems, and in these cases it is difficult to sustain the hypothesis of predictive models or learning processes. Furthermore, they are also observed in macromolecules and this excludes any possible explanation based on innate processes due to natural selection. Rosen concludes that a new law of causality is needed to explain the anticipatory behaviors of living systems.

Syntropy states that life depends on the future and that it continually manifests retrocausal behaviors of anticipation.

The hypothesis that living systems use a different type of causality had also been advanced by Hans Driesch (1867-1941), a pioneer in experimental research in embryology.

Driesch suggested the existence of final causes, which operate from the global to the analytic, from the future to the past. The final causes lead living matter to evolve towards the purpose of nature which Driesch called entelechies, from the Greek *en-telos* which means something that contains in itself its own purpose and that evolves towards this end. So, if the normal development path is interrupted, the system can reach the same end in another way. Driesch believed that the development and behavior of living systems were governed by a hierarchy of entelechies united by a single final entelechy.

Driesch provided the proof of this phenomenon by using sea urchin embryos. Dividing the sea urchin embryo cells after the first cell division, Driesch expected each cell to develop into the corresponding half of the animal for which it was designed or planned, but instead he discovered that each developed into a full sea urchin. This also happened in the four-cell stage: whole larvae developed from each of the four cells, although smaller than usual. It is possible to remove large pieces from the eggs, mix the blastomeres and interfere in many ways without affecting the embryo. It seems that every single monad in the original egg cell is able to form any part of the complete embryo. On the contrary, when two young embryos are joined, a single sea urchin is obtained and not two sea urchins.

These results show that sea urchins develop towards a morphological end. The moment we act on an embryo, the cell that survives continues to respond to the final cause that leads to the formation of structures. Although smaller, the structure that is reached is similar to the one that would have been obtained from the original embryo. It follows that the final form is not caused by the past or by a program, a project or a design that acts from the past, since any change we introduce in the past leads to the formation of the same structure. Even when a part of the system is removed or normal development is disturbed, the final form is reached which is always the same.

Another example is that of tissue regeneration. Driesch studied the process by which organisms are able to replace or repair damaged structures. Plants possess an extraordinary range of regenerative abilities, and the same happens with animals. For example, if a worm is cut into pieces, each piece regenerates a complete worm. Many vertebrates have an extraordinary capacity for regeneration, for example, if the lens of a newt's eye is surgically removed, a new lens is regenerated from the edge of the iris, while in the normal development of the embryo the lens is formed in a very different way, starting from the skin. Driesch used the concept of entelechy to explain the properties of integrity and directionality in the development and regeneration of living bodies and systems.

Independently in 1926 the Russian scientist Alexander Gurwitsch (1874-1954) and the Austrian biologist Paul Alfred Weiss (1898-1989) suggested the existence of a new causal factor, different from classical causality, which was called morphogenetic field. In addition to stating that morphogenetic fields play an important role in controlling morphogenesis (the development of body shape), the authors show that classical causality fails.

The term "field" is currently fashionable: gravitational field, electromagnetic field and morphogenetic field. It is used to indicate something that is observed, but is not yet understood in terms of classical causality; events that require a new type of explanation based on a new type of causality.

Syntropy replaces the terms "entelechies" and "fields" with the terms "final causes" and "attractors". Causes that act from the future produce fields that attract and guide.

Syntropy assumes that living systems are guided towards final causes by inner feelings that respond to attractors and that retrocausality is manifested mainly in the form of synchronicities. The same happens in our lives: inner feelings guide us towards the Attractor, the purpose of our existence.

A very important example was provided by Steve Jobs, the founder of Apple Computer.

Steve Jobs had been abandoned by his natural parents and this was the drama that accompanied him throughout his life. He was tormented and never accepted being abandoned.

He left university during the first year and ventured to India to find his inner self.

He discovered a completely different vision of the world that marked his change:

"in the Indian countryside people do not let themselves be guided by rationality, as we do, but by intuitions."

He discovered intuitions, a very powerful faculty, very developed in India, but practically unknown in the West.

He returned to the United States convinced that intuitions were more powerful than intellect. To cultivate intuitions it was necessary to live a minimalist life, reducing entropy as much as possible. He became a vegan, refused alcohol, tobacco and coffee, began to practice Zen meditation and had the courage not to be influenced by the judgment of others.

He always tried to reduce entropy to the point that it took him more than 8 months to choose the washing machine. He absolutely had to find the one with the lowest energy consumption and maximum efficiency. He lived in a thrifty way, a life so essential and austere that led his children to believe he was poor.

The way he lived was the result of his need to focus on the heart, on inner feelings. He avoided wealth because it could distract him from the voice of the heart. He was one of the richest men on the planet, but he lived like a poor man! From a syntropic perspective, his minimalist choices allowed intuitions to emerge, becoming the source of his innovations and wealth.

Jobs opposed marketing studies, as he said that people don't know the future. Only intuitive people can feel the future.

When he returned from India he saw an electronic board at his friend Steve Wozniak's house and he had the intuition of a computer that could be held in one hand. He asked Wozniak to develop a prototype of a personal computer, which he named Apple I. He managed to sell a few hundred and this sudden success gave him the impetus to develop a more advanced model, suitable for ordinary people, which he called Apple II.

Jobs was not an engineer, he had no scientific or technical mind, he was simply an artist! What do computers have to do with his life? Jobs had nothing to do with electronics, but his intuitive abilities showed him a goal, an object of the future. Thirty years earlier, in 1977, he had sensed a pocket computer that combines aesthetics, simplicity, technology and minimalism! He felt the need for a product that, in addition to being technologically perfect, had to be also beautiful and simple!

His obsession with beauty and simplicity led him to devote an enormous amount of time to the details of the Apple II. It had to be beautiful, silent and at the same time essential and simple! It was an unprecedented commercial success that made Apple Computer one of the leading global companies.

Jobs noticed that when the heart gave him an intuition, it turned into a command he had to follow, regardless of the opinions of others. The only thing that mattered was finding a way to give shape to the intuition.

For Jobs, the vegan diet, Zen meditation, a life immersed in nature, abstention from alcohol and coffee were necessary to nourish his inner voice, the voice of his heart and strengthen his ability to intuit the future.

At the same time, this caused great difficulties. He was sensitive, intuitive, irrational and nervous. He was aware of the limitations that his irrationality caused him in handling a large company, such as Apple Computer, and chose a rationalist manager to run the company: John Sculley, a famous manager he admired but with whom he entered continually in conflict, to the point that in 1985 the board of directors decided to fire Jobs from Apple, the company he had founded.

Apple Computer continued to make money for a while with the products designed by Jobs, but after a few years the decline began and in the mid-1990s it came to the brink of bankruptcy. On December 21, 1996, the board of directors asked Jobs to return as the president's personal advisor. Jobs accepted. He asked for a salary of one dollar a year in exchange for the guarantee that his insights, even if crazy, were accepted unconditionally. In a few months he revolutionized the products and on September 16, 1997 he became interim CEO.

Apple Computer resurrected in less than a year. How did he do it?

He said we should not let the noise of others' opinions dull our inner voice. And, more importantly, he repeated that we must always have the courage to believe in our heart and in our intuitions, because they already know the future and know where we need to go. For Jobs, everything else was secondary.

Being interim has marked all his new products. Their name had to be preceded by the letter *i*: *i*Pod, *i*Pad, *i*Phone and *i*Mac.

Jobs's children believed he was poor. They often asked him:

"Daddy, why don't you take us to one of your rich friends?"

He talked about important business walking in parks or in nature. To celebrate a success, he invited employees to restaurants for $10 per person. When he made a gift he collected flowers in a field. He wore the same clothes for years. Despite the immense riches he had!

He was convinced that money was not his, but that it was a tool to reach the end.

At the time of Apple I, he repeated that his mission was to develop a computer that could be held in one hand and not to get rich. For him money was exclusively a tool.

The ability to feel the future was the source of Jobs' wealth. It was the ingredient of his creativity, genius and innovation.

Einstein repeated that: "*the intuitive mind is a sacred gift and the rational mind is its faithful servant. But we have created a society that honors the servant and has forgotten the gift.*"

Zen meditation helped Jobs calm his mind and move the attention to the heart.

In his lectures he used to say that almost everything, expectations, pride and fears of failure, vanish in the face of death. He emphasized the centrality of death and the fact that when we are aware of dying we pay attention only to what is really important. Being constantly aware that we are destined to die is one of the most effective ways to understand what is really important and to avoid the trap of attaching ourselves to materiality and appearance. We are already naked in the face of death. Since we must die, there is no reason not to follow our heart and do what we have to do.

Jobs believed in the invisible and in synchronicities. He built the headquarters of Pixar (one of his companies) around a central space, a large square where everyone had to go through or stop if they wanted to eat something or use the services. In this way the invisible world was favored by chance encounters.

According to Jobs, chance does not exist.

Chance encounters allow the invisible, to activate intuitions, creativity and synchronicities and make visible what is not yet visible.

Jobs loved to quote Michelangelo's famous phrase:

"In each block of marble I see a statue as if it were in front of me, shaped and perfect in attitude and action. I just have to remove the rough walls that imprison the beautiful appearance to reveal it to others as my eyes see it."

Jobs believed that we all have a task, a mission to carry out. We just need to discover this mission by removing what is not necessary.

Jobs made visible what he had intuited. He died a few months after the presentation of the *i*Pad, the computer that can be held in one hand, the mission of his life.

The life of Jobs testifies that intelligence and creativity come from the future, from the invisible and that we can access the invisible through intuitions.

He showed that the voice of the heart brings the future into the present. Rainer Maria Rilke said: *"The future enters us, to become us, long before it happens."*

Now let's move on to another example.

The complementarity between entropy and syntropy can be represented as a seesaw with causality on one side and retrocausality on the other.

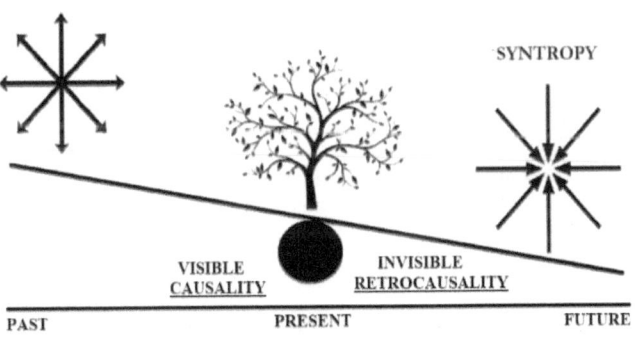

Life is the manifestation on the physical plane of syntropy. It is constantly in conflict with entropy and must always diminish it. However, this is hampered by our activities that tend to increase entropy.

The challenge of life is:

how to increase syntropy
and reduce entropy
by remaining active?

To describe this challenge I will use the example of a freelance, single, whose expenses exceeded the income of over five hundred euros a month.

The savings were running out and he had no one to ask for help. He started reducing his expenses: no money in his wallet, no credit on his cell phone. But things went from bad to worse. At this point he asked me for help.

Let's see how it went:

«How much do you spend on your mobile phone?»

«About 40 euros a month, but I always find myself without credit.»

«Why don't you change provider? There are interesting promotions. With only 10 euros a month you can have unlimited minutes and SMS and 20 gigabytes of internet.»

Lowering entropy means saving, but this must be done by maintaining or increasing the quality of life. For example, by changing an old contract. In this case, changing provider and choosing a new contract has led to an increase in the quality of life and to save over three hundred euros a year!

The trick is to improve the quality of life by saving.

When entropy (expenses) and syntropy (incomes) are balanced, the invisible world begins to manifest.

In this example we need to reduce spending by at least six thousand euros a year.

«Do you take shirts to the laundry to be ironed?»

«I wash them, but I am not able to iron them. I take them to the laundry to have them ironed.»

«How much does it cost you?»

«Between 50 and 70 euros a month.»

«Why don't you ask your maid if she can iron them for 8 euros more per month?»

The maid immediately accepted. Another small optimization that led to save over six hundred euros a year, but which significantly increased the quality of life by eliminating the hassle of going to the laundry. Again an increase in the quality of life while saving! These first two optimizations reduced entropy by around one thousand euros a year and increased the quality of life. The goal is to reach six thousand euros to balance income and expenses.

«*Do you go to work by car?*»

«*I also use the scooter to save money, but the traffic is really dangerous!*»

«*Why don't you use your bicycle?*»

«*On these roads ?!*»

«*No, on alternative roads.*»

«*My house is in the city center, the office is not far away, but I have always considered the bicycle impossible due to the difference in altitude of over 30 meters. I would arrive tired and sweaty.*»

«*If you have to climb it is better to choose a steep but short road, get off and push, rather than pedaling.*»

Thus he discovered the beauty of the

streets of the city center and parks. In less than 25 minutes he could reach his office by bicycle. It took more time by car or scooter. The next day he sold the scooter, canceled the insurance and the garage. In total, another three thousand euros saved per year. With this simple optimization, he has received other advantages: he exercises and no longer needs to go to the gym, more money and time saved! Moreover, he spends less on fuel.

Entropy has now decreased by over four thousand euros a year and the quality of life has improved!

We need to find another two thousand euros before syntropy, the invisible world can begin to show.

«*Your electricity bill exceeds 200 euros every two months! As a single you should not pay more than 50 euros.*»

«*What should I do?*»

«*Try using low energy light bulbs, such as LED lamps, and set the timer to the water heater.*»

Small changes that required little time and

money. One hundred and fifty euros saved every two months, nine hundred euros a year. With this small optimization he felt consistent with his ecological beliefs and the quality of life increased. Now he had reduced his expenses by over five thousand euros a year! We must reach the goal of six thousand euros a year!

«How much do you pay for electricity in your office?»
«About 300 euros every two months.»
«Do you use halogen bulbs !?»
«Yes.»

He discovered that he could save over a thousand euros a year simply by replacing the halogen spotlights with LED spotlights.

Now that the expenses no longer exceed the incomes, syntropy can begin to show itself in the form of synchronicities: meaningful coincidences.

Jung and Pauli have coined the term synchronicity to indicate an invisible causality different from that familiar to us. Synchronicities manifest as meaningful coincidences, because they converge towards an end.

Invisible causality acts from the future and groups events according to purpose. Synchronicities are significant because they have a purpose.

«How much do you pay for renting your office?»

«Nothing. It is owned by my aunts.»

«They could rent it and make a profit, but you use it for free ?!»

«Exactly.»

«And what are your aunts living on?»

«They both receive a pension and have some savings, but their financial situation is not good, they constantly complain.»

«Have you ever thought about renting a room in an office and letting your aunts rent their apartment?»

«I have no money, I can't afford to pay a rent!»

«How's your business going?»

«I have few clients, perhaps because of the economic crisis, but also because of the position of the office.»

«A less prestigious office, but in a strategic and

well-connected place could help you have more customers ?!»

The first synchronicity is the following. The day after this dialogue, as if by magic, he received the offer of a room in an office in the most central area of the city, at the price of only 250 euros a month, including all utilities! The aunts' apartment was in a very beautiful and prestigious place, but difficult to reach and there was no parking: beautiful, prestigious, but inconvenient and very expensive. However he hesitated, he didn't dare!

The next day another synchronicity occurred. He received a call from the doorkeeper. An airline offered 2,800 euros a month for his aunts' apartment. Obviously the aunts asked him to find another place immediately and fortunately the day before he had received the offer of a room. But he still wasn't convinced. The office in the city center was in a very noisy area: well connected, but chaotic.

The third synchronicity is the following.

That same afternoon he was walking in the area of the city he likes most. It is not central, but it is green, quiet and well connected. At a shoemaker's window, he saw a notice for a room in an office. The apartment was in the building next to the shoemaker. He called and immediately went to see it. He instantly decided to rent the room. In a city like Rome it is difficult to find rooms for rent in professional studios and above all in such a beautiful place of the city.

When synchronicities are activated, we are attracted to places and situations that otherwise we would not have taken into consideration and that solve our problems. Synchronicities are accompanied by feelings of warmth and well-being in the thoracic area that inform us that we are on the right path.

«I began to feel warmth and well-being in the chest area. My clients like the new studio. There is a parking lot, it is nice, quiet and it is located near a metro station. My business is thriving, my savings are increasing and my personal and sentimental life has improved.»

Syntropy offers wealth and happiness. But when things go well it is easy to fall back into the old entropic and dissipative lifestyles.

A few months later he received a job offer, a prestigious job abroad: his dream!

He immediately accepted and moved. The salary was high, taxation was low. Suddenly he would become a rich man who could lead the rich life he had always wanted.

But this reverses the balance between entropy and syntropy: wealth leads to living in an entropic way, entropy increases and syntropy decreases and we go back to failure!

«The foreign company was only interested in making money, without any ethics. I had to work almost fifty hours a week, there was nothing else outside the company. It was necessary to give absolute priority to what was profitable, even if immoral. A few months later I felt disgusted with my profession. Taxes were low, but I had to pay all the services. By adding the rent of the house and the expenses related to the fact that I was a foreigner, I paid much more than I earned. After only six months I had accumulated

more than twenty-eight thousand euros of debts! The dream had broken and had become a nightmare. From heaven I fell to hell. I had no time for myself or for my love life. First I felt discomfort, then suffering, and eventually depression and anxiety exploded. I decided to go back to Italy!»

This often happens. Syntropy increases the quality of life, well-being, but also wealth. As soon as material wealth returns people fall into entropic lifestyles.

For this reason the increase in syntropy must be accompanied by an inner transformation. People do not have to consider money as their property, but as a tool. They must be aware that happiness is not achieved through wealth, but thanks to the fulfillment of our mission.

If this inner transformation is lacking, the process fails.

Material improvements must be accompanied by a new awareness of the invisible.

Wealth is only one aspect of the game between entropy and syntropy. When wealth is reached without an inner transformation it is inevitable to fall back into entropy and suffering.

This game between entropy and syntropy involves not only individuals, but also companies, institutions and nations. It can be used successfully in the management of cities, nations, public and private organizations and ecological and natural systems. But it must always be accompanied by an inner transformation that puts the heart at the center of decision-making, otherwise it will inevitably lead to failure.

The compass of the heart is of great importance in the game of life, but since in the same area we perceive emotions linked to fear and danger it is not easy to use.

These emotions are activated by the amygdala.

The amygdala is designed to ensure survival. When we are faced with a danger it releases hormones that trigger the fight or flee reaction. The amygdala is fast, but inflexible. The emotional charge enters our body and covers the feelings of the heart.

Fears and dangers limit the ability to use the compass of the heart and increase entropy.

The compass of the heart requires that we silence fear and the chatter of the mind.

A very effective way is provided by Zen meditation.

During Zen meditation participants cannot react to stimuli, but they can only observe them. Practicing Zen meditation we discover that thoughts wait for the reaction of the heart. When the heart reacts it provides energy to the thought which becomes stronger. When we don't react the thought dissolves.

The heart decides when to react and to be silent; the mind can only adjust to the will of the heart. We are the heart. Our will is in the heart. In this way the scepter of command moves from the head to the heart and the mind becomes silent.

The importance of silence can be found in many traditions. The groups of Friends (also known as Quakers) started practicing silence in 1650 when George Fox discovered that it restores the flow of the energy and a direct contact with Deity. The practice is simple, people sit in a circle and are silent for about an hour. Shared silence helps to feel the heart.

Silence is a natural technique, a simple and enjoyable way of being together with others. It is not a religion and does not require devotion to a faith, or to a specific philosophy. It creates distance from our thoughts. It frees our being from the conditioning power of the words and leads to discover that we are part of something broader. When the chatter of the mind ends we experience a new condition: to be without thinking. A state in which thoughts are produced only when required by the heart. A state in which the gap between a thought and the other is not empty, but it is pure and absolute potentiality. Being without thinking empowers the heart: our true will.

Another factor which influences the perception of the heart is what we eat.

John Hubert Brocklesby became a vegetarian in prison during the First World War. For him, Christians did not have to kill other Christians and declared himself a conscientious objector. He was arrested and imprisoned in the Richmond Castle. He had to face court martial. He knew he would be sentenced to death and he was terrified at the idea.

Another conscientious objector told him: «*If you talk with your heart it is God who speaks through you.*» This gave him courage. Then this conscientious objector added: «*If you do not eat meat, the voice of the heart becomes stronger.*»

John Hubert Brocklesby became a vegetarian in prison to serve the will of God and face court martial.

A book was written using his diaries.[34]

Since we have a vegetarian structure (no claws to hunt, teeth suitable for fruit and a long digestive system not meant for meat) the attractor towards which we are converging has these features. Therefore, being vegetarian helps the connection with the attractor, increasing the flow of syntropy and the feelings of the heart.

This last consideration is supported by an epidemiological study conducted by the Canadian Natural Hygiene Society on the risk

[34] Jones WE, *We Will Not Fight: The Untold Story of World War Ones Conscientious Objectors*, www.amazon.com/dp/1845133005/

of heart attacks that shows that meat eaters have a 50% risk, vegetarians 15%, vegans 4%.

Among the diet options that seem to increase the perception of the heart one is liquidarism.

Michael Werner, born in 1949 in northern Germany and CEO of a pharmaceutical research institute in Arlesheim, became liquidarian in January 2001 and since then drinks only water and does not eat solid food. In his book *Living on Light* Werner says that:

"I found that my conversion to living without food went extraordinarily well. I expected to feel weaker and weaker during the first few days. But then I began to realize that in my case this weakness did not exist. Instead I experienced a growing feeling of lightness during the day and a decrease in the amount of sleep I needed during the night. Going through this process was probably the most intense experience of my adult life."

If it is true that one can live and be fit and healthy without eating, incredible scenarios open up about human life and life in general.

Werner notes that being liquidarian is different from fasting:

"It is something completely different! With fasting the body mobilizes reserves of energy and matter and one cannot fast for an unlimited time, nor can one be without drinking. But the process I was undertaking was and remains a mental-spiritual phenomenon that requires a particular inner predisposition. In reality there is a condition: opening up to the idea of being able to be nourished by the etheric, by prana or by whatever it may be called. This is the necessary requirement. Then it will happen. I live liquidarism as a gift from the spiritual world."

Rudolf Steiner (1861-1925), Austrian philosopher, social reformer, architect and esotericist, attempted to formulate a spiritual science, a synthesis between science and spirituality that applied the clarity of scientific thought, of Western philosophy, to the spiritual world. Steiner believed that matter was condensed light (he used the word light with the same meaning of syntropy). If matter is condensed syntropy, there must be many different ways to transform the invisible (syntropy) into matter. Our visible environment is immersed in an invisible environment, a syntropic reality that offers incredible possibilities, including that of living from syntropy. Steiner believed that life was impossible without syntropy (ie without light), since syntropy is the vital energy that we continuously and directly absorb. To live only on water it is necessary to believe that it is possible to "*live by syntropy.*" According to Steiner, the act of digesting stimulates the body to absorb the vital energy from the invisible, which is transformed and condensed into substance that maintains and builds our body. Steiner used the following example: when we eat a potato, we chew and digest and this leads to absorbing the vital forces from our etheric environment and condensing them into substances. In other words, our body acquires structure and substance absorbing syntropy and invisible forces.

Michael Werner emphasizes that the only prerequisite for feeding on light (ie syntropy) is to trust it. He uses the words of Steiner:

"There is a fundamental essence of our earthly material existence from which all matter is produced through a process of condensation. What is the fundamental substance of our terrestrial existence? Spiritual science gives this answer: every substance on earth is condensed light! There is nothing but condensed light ... Wherever you touch a substance, there you have condensed light. All matter is, in essence, light."

In other words, all matter is nothing else but condensed syntropy!

But, it is important to be careful. Many people suggest fasting, nonetheless some techniques can be dangerous, as it is the case of Jasmuheen's breatharianism, a fast without food and liquids that has been lethal to various followers.[35]

---

[35] Di Corpo U., Liquidarism, Syntropy and Vital Needs: https://www.amazon.com/dp/1092909060

VITAL NEEDS
AND
THE INVISIBLE FORCE OF LOVE

Water is the lifeblood that provides syntropy to life. Without water life is unable to counteract the destructive effects of entropy and dies. We can therefore list water among the vital needs.

Life also needs energy. This is why the Sun is so important. The chlorophyll process absorbs energy from the Sun and without the Sun life could not exist on this planet.

Life dies when water freezes. Heat is needed to keep life away from low temperatures.

Living systems are generally not able to feed directly on syntropy. Therefore they must meet conditions for the acquisition of food. These conditions are known as material needs.

When these needs are not met, alarm bells are activated, such as thirst for the need for water, hunger for food and chill for the need for heat.

These alarm bells are well known to all, we know how to associate them with the need that must be met and we know what we must do.

But we also have invisible vital needs !!!

The *Attractor* is the source of syntropy and resides outside of our physical body, connected to it through the solar plexus. It provides visions of the future, insights, inspirations and higher levels of awareness, which are inaccessible to the ordinary states of the rational mind. It shows the direction, the goals and the mission of our life by acting as a teacher that guides us to the solution of problems and to well-being.

We establish the connection through the autonomic nervous system, the solar plexus, which we commonly associate with the heart.

This connection is easier and stronger in moments of meditation and love and when we abstain from the consumption of alcohol, tobacco, drugs and coffee and when we follow a vegetarian or liquidarian diet.

Since syntropy concentrates energy, a good connection is perceived as warmth and well-being in the solar plexus. On the contrary, a weak connection is signaled by feelings of emptiness and pain that we usually indicate as anxiety and by symptoms of the autonomic nervous system, such as nausea, dizziness and suffocation.

Syntropy is needed to regenerate damaged cells and parts of the organism. The autonomic nervous system acts like a mechanic who consults the manufacturer's guide to carry out repairs and keeping the system as close as possible to the design. However, the design is not mechanical and the instructions are written with the ink of love.

The autonomic nervous system is in charge of all the involuntary functions of the body and is responsible for controlling the movement of muscles and limbs and regulates body functions that are not subject to decisions and that do not require the conscious mind. For example, it is responsible for digestion, heart rate, food assimilation and cells regeneration.

These processes are completely unknown to our conscious mind. We don't know how they are performed and often we don't even know they exist. We don't need to be a doctor or a biologist to digest food or regenerate tissues. The body knows everything and shows an extraordinary level of intelligence. It directs and regulates these processes, thus expressing the capacity and potentials of an intelligence that is incredibly superior to our conscious mind.

It develops patterns of behavior that it then performs autonomously and automatically and that are maintained over time, giving rise to habits that are then stored, at least in part, in the muscles of the body. Behavioral patterns are repeated until they are activated automatically, regardless of our will. These patterns are then firmly placed in the memory of the unconscious mind. The conscious mind often does not know what is in the memory of the unconscious mind. As a result, the unconscious mind can open incredible scenarios in the processes of knowing ourselves. The autonomic nervous system (ie the unconscious mind) also acts as a guardian of any information that the conscious mind cannot handle.

When the connection with the attractor is strong we feel warmth, well-being and love, when it is weak we feel void, pain and anxiety accompanied by loneliness and isolation. In the absence of the connection the autonomic nervous system is not able to provide syntropy to the vital functions and the organism dies.

We can therefore die not only because of unsatisfied material needs, but also because of the lack of connection with the attractor.

The need for connection with the attractor is usually perceived as a *need for love* and cohesion.

To respond to our needs, we build maps of the physical environment that lead to realize that we live in a world that has expanded towards infinity. On the contrary, consciousness concentrates towards the infinitely small.

The identity conflict arises from the comparison:

$$\frac{I}{Outside\ World} = 0$$

When I compare myself to the outside world I am equal to zero

By comparing ourselves with the physical reality we realize that we are equal to zero and this is incompatible with the feeling we exist.

This conflict is well described in Shakespeare's Hamlet with the phrase "*to be or not to be*". Not being is incompatible with life. To continue to respond to the challenges of life we need to find a purpose, a meaning, otherwise it is all useless.

The identity conflict leads to a vital **need for meaning** which, when not satisfied, causes feelings of worthlessness and depression.

Depression is an unsustainable type of suffering and people face it trying to inflate their Ego, limiting the size of the world they are comparing to or simply erasing the outside world.

However we manipulate the numerator and/or the denominator of the equation of the identity conflict the result continues to be always equal to zero.

The need for meaning is an invisible need. Most people are not aware of it, but still it is vital and we must constantly respond to it.

We must all give meaning to our life and to do so we often accept the most incredible contradictions.

The identity conflict equation suggests a solution:

$$\frac{I \times Outside\ World}{Outside\ World} = I$$

When I compare myself to the outside world

This is called the *Theorem of Love* and shows that:

- only when our inner world unites with the outside world through love, we overcome the identity conflict;
- love provides this unity (I x Outside World), and therefore love is vital: it gives meaning to life;
- love allows to shift from duality (I = 0) to unity (I = I).

When we love, we converge towards unity and our heart is filled with warmth, well-being and happiness. When we do not love, we diverge and we experience pain, emptiness and loneliness and our life is meaningless.

Today the word love is abused and can mean anything! So let's see how it is used in this book.

First of all, love it is something that we feel in the form of warmth and well-being in the thoracic area. It may be accompanied by an increased heart rate, sweating, shortened breath, redness, dilated pupils.

Love is vital because it gives meaning to life and because it connects us with the Attractor.

What activates love becomes vital. For this reason, when we find a source of love we tend to cling to it and forget everything else. In the absence of love, suffering can become unbearable.

Let's recap:

– The first group of vital needs is commonly known as ***material needs***. To combat the dissipative effects of entropy, living systems must acquire syntropy through water, energy and food, they must protect themselves from the dissipative effects of entropy and eliminate the remains of the destruction of structures by entropy. These conditions include shelter, clothing, waste disposal and hygiene. The partial satisfaction of material needs is signaled by hunger, thirst and various forms of suffering. Total dissatisfaction leads to death.

— The second vital need is commonly called the **need for love**. Responding to material needs does not prevent entropy from destroying life. For example, cells die and must be replaced. To repair the damage caused by entropy, we must draw on the regenerative properties of syntropy which allow to create order, reconstruct structures and increase the levels of organization. The autonomic nervous system, which supports vital functions, acquires syntropy. Since syntropy acts as an absorber and energy concentrator, the intake of syntropy is felt in the thoracic area of the autonomic nervous system, in the form of warmth and well-being that we usually indicate as love; the lack of syntropy is perceived as emptiness and pain in the thoracic area, usually referred to as anxiety. In short, the need to acquire syntropy is felt as a need for love. When this need is not satisfied there is suffering in the form of emptiness and pain. When this need is totally unsatisfied, living systems are not able to sustain the regenerative and vital processes and entropy takes over, bringing the system to death.

— The third vital need is commonly called the ***need for meaning***. In order to satisfy material needs we produce maps of the environment. These maps give rise to the identity conflict. Entropy has inflated the physical universe towards infinity, while syntropy concentrates consciousness in extremely limited spaces. As a result, when we compare ourselves to the infinity of the universe, we discover that we are equal to zero. On the one hand we feel we exist, on the other we are aware of being equal to zero. These two opposing considerations *"to be or not to be"* cannot coexist. The identity conflict is characterized by lack of meaning, lack of energy, existential crisis and depression, generally perceived in the form of tensions in the head accompanied by anxiety. Being equal to zero is equivalent to death, which is incompatible with our feeling of existing. From this arises a vital need for meaning.

The solution to suffering is provided by the Theorem of Love. The Theorem of Love requires that we rely on the heart (the solar plexus) and use it consciously and intentionally to go towards the most beneficial options.

Love is an invisible force, an inner power within, which provides enthusiasm. In Greek enthusiasm means *"God within"*, an invisible force which lets us overcome the most incredible difficulties and endeavors.

The metaphor of the cart can help summarize. In this metaphor:

- the cart is the physical body and requires maintenance;
- the horses are our impulses, that pull us in different directions and give the movement; they require energy and the guide of the coachman;

- the coachman is the mind, follows the orders of the master, directs the horses and takes care of the cart;
- the master of the cart is the heart which provides direction and aim.

All functions well when:

- The cart is well cared for (*material needs*).
- Horses receive water and energy.
- The coachman follows the directives of the heart (ie the master).
- The master is guided by Love, by the Attractor. Love provides aim and objectives.

ATTRACTORS

The energy-momentum-mass equation suggests that the present can be described as the meeting point of causes that act from the past (causality) and attractors that act from the future (retrocausality).

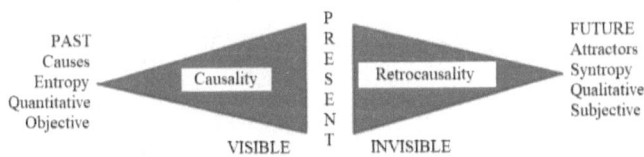

Causality requires a big cause for a great effect. This is due to the fact that causality diverges and tends to disperse. On the contrary with attractors the effect is amplified. The smaller the cause, the more it can be amplified and the greater the effect.

This strangeness of attractors was discovered in 1963 by the meteorologist Edward Lorenz. When dealing with water, as happens in meteorology, a small variation can produce an amplifying effect. Lorenz described this situation with the famous phrase:

"The flap of a butterfly in the Amazon can cause a hurricane in the United States".

However, for this to happen it is necessary that the small flap (the active principle) is in line with the attractor. Otherwise entropy prevails and the small energy of the flap is lost. On the contrary, when the variation is in line with the attractor it is amplified.

The hydrogen bond of water operates in both directions: from the micro to the macro, amplifying the effect, and from the macro to the micro informing the attractor. This can help understand how homeopathic remedies work.

Homeopathy is based on water. When we insert into water the similar, the *simillimum*, of what we want to cure, its information enters the quantum level and informs the attractor. The greater the dilution, the greater the contribution of the attractor in the amplification of the effect.

Homeopathy is the subject of ferocious attacks. In Italy the famous scientific television journalist Piero Angela reiterates that "*homeopathy is fresh water*", "*pseudoscience*" or even "*magical practice*" and constantly emphasizes that it has no scientific validity. "*It is a placebo effect, this is what the scientific community says.*" Angela underlines that "*for Rita Levi Montalcini (Italian Nobel Prize) homeopathy is potentially harmful because it distracts patients from valid treatments*" and that "*for Renato Dulbecco (another Italian Nobel Prize) it is a practice without any value.*" Lately the attacks on homeopathy have intensified; the main accusations are that homeopathy is only fresh water and a placebo effect.

Experimental studies show the effectiveness of homeopathy, but conventional medicine continues to consider homeopathy non-scientific since the "active substance" (the solid substance) has been completely removed from water by dilution. It is considered impossible that water can be the cause of the effects observed in the experiments, since it is considered an inert substance.

Homeopathy was discovered in 1796 by the German doctor Samuel Hahnemann (1755-1843). This system is based on the so-called law of similes, according to which the remedies must use substances that cause similar symptoms in healthy individuals. These substances are then diluted in water. The strange fact is that the higher the dilution the more powerful is the effect. The most powerful remedies are those in which the substances have been diluted to the point that it is impossible for a single molecule to still be present in the remedy. For conventional medicine, after removing the active ingredient through dilution, effects can only be placebo effects, not attributable to the remedy, since no solid molecule of the active ingredient is present.

Syntropy claims that the active ingredient, when placed in water, creates links with attractors. So by removing the active ingredient through dilution, these retrocausal bonds remain and are no longer related to the substance but are free to act on any other structure.

Syntropy explains the effects of homeopathy as a consequence of the retrocausal properties of water.[36] The remedies act from the future and the effects are the result of the interaction between causality that is governed by entropy and retrocausality that is governed by syntropy.

When using a substance that induces in the future of a healthy person symptoms similar to those observed in a sick person and this substance is diluted in water (beyond the value of Avogadro), the future begins to retroact into the present.

With causality in order to increase the effect it is necessary to increase the cause (the active substance), while with retrocausality in order to increase the effect it is necessary to reduce the cause. Retrocausality works in the opposite way to causality. This explains why in homeopathy to enhance the remedy instead of increasing the active substance this is diluted.

---

[36] Paolella M.., *Homeopathic Medicine and Syntropy*: http://www.sintropia.it/journal/english/2014-eng-2-01.pdf

Homeopathy cannot be explained on the basis of classical causality, since the active ingredient is completely removed from homeopathic preparations (which are water based). The therapeutic effects, however, are obvious and can be demonstrated experimentally. The results are strong even when no placebo effect is possible, as in the case of studies carried out on plants in agriculture.

The retrocausal properties of water are due to the hydrogen bond. The hydrogen atoms are in an intermediate position between the subatomic (quantum) and the molecular level and provide a bridge that allows syntropy to flow from the attractor to the macroscopic level.

One of the objections to evolution by random mutations is the fact that the simplest proteins are made of chains of 90 amino acids and that combinatory calculations show that more than 10^{600} permutations (ie one followed by 600 zeros) are necessary to randomly combine amino acids into a "spontaneous" protein of 90 amino acids.

Walter Elsasser, in a work published in the American Scientist[37], shows that in the 13-15 billion years of our Universe no more than 10^{106} events took place (also considering the level of nanoseconds). Consequently, any event requiring a combinatorial value greater than 10^{106} is simply impossible in our Universe.

The number 10^{600} is by far greater than all the possible combinations in the history of our Universe. In other words, the possibility that only one protein is formed by chance is null.

Elsasser's results show that: "*the notion of chance in biology has no logical foundation ... its use to explain life is at best metaphorical, but there is a danger that this metaphor may divert attention in the wrong direction.*"

Life shows an incredible complexity that converges towards common projects, despite

---

[37] Elsasser W.M., *A causal phenomena in physics and biology: A case for reconstruction.* American Scientist 1969, 57: 502-16.

individual differences. For example, we can recognize different races, such as Europeans, Asians, Africans, but there is something that unites all these individuals and that makes them all human beings.

Considering only the contribution of the past, it is impossible to explain why individuals converge towards common projects and it is impossible to explain the stability of these projects over time.

Attractors describe this stability and this convergence.

The biologist Rupert Sheldrake has devised experiments that show that when individuals of the same species learn to solve a task, this knowledge spreads invisibly and immaterially to all the other individuals of the same species.

Attractors behave like relays. When an individual solves a task and receives a benefit, the information is relayed to all the other individuals.

Attractors establish a bridge between individuals that allows them to develop a shared knowledge.

Individuals converging towards the same attractor are able to share knowledge invisibly, without the involvement of any physical means. This is known in quantum mechanics as entanglement and non-locality.

Attractors receive information and experiences from individuals, select what is advantageous and redistribute it. This process transforms individual experiences into intelligent information, which provides solutions, projects and form.

The verb "to inform" comes from the Latin "in-formare", which means "to give form". Aristotle believed that "in-formation" was a fundamental activity of energy and matter. In-formation does not have an immediate meaning, like the word "knowledge", but rather implies a modality that leads to the creation of forms. Once a form takes shape, it can be manifested in all individuals who are connected to the same attractor.

People often ask if attractors imply that the future is already determined. The answer is simply NO, they imply exactly the opposite!

Attractors indicate that we will inevitably return to where syntropy originates, to the attractor, what Teilhard de Chardin calls the *Omega point*, but that the path depends on our choices.

If attractors did not exist, we would live in a mechanical universe totally determined by the past. Instead we are constantly forced to choose between the head and the heart, between past and future.

Water is not an inert liquid, it is the means by which we connect with the attractor, information and nourish the vital processes of the body. The hydrogen bond provides water with properties different from those of all other liquids. These properties explain a wide range of phenomena that medicine is not yet able to accept.

Water provides syntropy to living organisms and when there is a lack of water, entropy prevails, causing suffering and symptoms that are often interpreted by conventional medicine as organic diseases.

In the book *"Your Body's many Cries for Water"* the Iranian doctor Fereydoon Batmanghelidj (1931-2004) offers an important explanation of the role of water in life, and specifically in the human body.

Batmanghelidj completed his medical studies at St. Mary's Hospital in London and opened several clinics when he returned to Iran. However, during the 1979 Iranian revolution he was arrested and spent almost three years in prison in Tehran. A prison that was designed for 600 people, but which housed more than 9 thousand people.

Here is how Batmanghelidj describes his discovery:

"The nightmare of life and death in that hell hole threatened everyone and tested the courage and strength of the weak and the strong. It was then that the human body revealed to me some of its greatest secrets, secrets never understood by medical science. (...) One night, after about two months of imprisonment, that secret was revealed. It was about 11 pm. I woke up, one of my cell mates suffered from terrible stomach pains. He couldn't walk alone. Others were helping

him stand up. He suffered from peptic ulcer and needed medical attention. He was very ill, but I was not allowed to take any medicine with me. At this point the surprising event occurred! I gave him two glasses of water and the pain disappeared within minutes and he could stand on his own again."[38]

Due to extreme conditions in Tehran prison, Batmanghelidj was able to discover that many diseases can be healed simply with water. Batmanghelidj came to the conclusion that the lack of water is expressed not only by thirst and dry mouth, but also by a series of localized symptoms that serve to inform us about a local need for water. These local signs of dehydration take the form of pain and are usually interpreted as symptoms of illness and not the need for water. Batmanghelidj realized that we often mistake pains caused by a local dehydration situation for diseases.

Conventional medicine concentrates on the solid 25% and does not consider the role of water (ie the other 75% of the body), since it

---

[38] Batmanghelidj F (1992), *Your Body's many Cries for Water,* www.watercure.com

assumes that the solid part is the active principle and that all the functions of the body depend on the solid while water works only as a solvent that fills the space.

The human body is considered as a large "test tube" filled with different types of solids and water as a chemically inert and insignificant packaging material.

Conventional medicine assumes that solutes (substances dissolved or transported in the blood) regulate all the activities of the body, while it is assumed that the intake of water (the solvent) is generally well respected, since water is easily available.

Based on this hypothesis, medical research has been addressed to the study of solids that are considered responsible for the onset of diseases. To date, a dry mouth is the only recognized symptom of dehydration. However, according to Batmanghelidj, a dry mouth is only the ultimate symptom of extreme dehydration.

Dr. Batmanghelidj explains several diseases as a result of water deficiency: rheumatoid arthritis, hypertension, high cholesterol, excess body weight, asthma and some allergies.

According to Batmanghelidj the fundamental error of conventional medicine is to confuse dehydration with disease. This error inhibits the necessary preventive measures and the patient is not provided with sufficient water treatments to cure his suffering. At the first appearance of pain, the body should receive water. In contrast, conventional medicine provides drugs that block the symptoms of the lack of water and the consequent conversion of symptoms into chronic diseases and chronic dehydration.

Batmanghelidj suggests changing the medical paradigm, moving from a vision centered on the properties of the solute (solid matter ie past causes) to a vision centered on the properties of the solvent (water ie attractors).

Batmanghelidj states that the solvent (water) regulates the functions of the body, including the activities of all solutes (solids) dissolved in it.

In this new paradigm diseases are interpreted as disorders of the body's water metabolism (solvent metabolism).

Water carries nutrients, hormones and chemical messages and performs multiple vital functions. The balance between chemical and solid substances is restored by restoring the correct water balance. In light of these considerations, water becomes the natural cure for a wide spectrum of disorders and complications that are currently labeled as "diseases".

Attractors bring parts together. The unity of our Self is strengthened when we have a mission, when we are converging towards an attractor. When, on the contrary, we have no attractor cohesion diminishes, the chatter of the mind increases and our personality shatters.

Converging is therapeutic since it brings together our parts and makes them cooperate.

The evolutionary paleontologist Teilhard de Chardin noticed that the incredible stability of species is given by the fact that they converge towards attractors. He advocated the idea that life is guided by attractors, and evolves according to a hierarchy of attractors, till the ultimate unifying attractor, the Omega point, is reached.

Since they reinforce the Self, attractors increase individualization, nonetheless they also lead towards unity.

It seems a contradiction, but unity and diversity go together!

The theme of attraction has been the focus of Teilhard's research:

"Reduced to its essence the problem of life can be expressed like this: accepting the two principles of conservation of energy and entropy, how can they assimilate without contradiction, a third universal law (which is expressed by biology), that of the organization of energy? ... the situation becomes clear when we consider, at the basis of cosmology, the existence of a sort of anti-entropy."

Teilhard formulated the hypothesis of a converging energy, similar to what Fantappiè discovered with syntropy.

"*In other words, not just one kind of energy, but two different energies; two energies which cannot transform directly one into the other, because they operate at different levels ... The behaviour of these two energies are so completely different and their manifestations so completely irreducible that we might believe they belong to two completely independent ways of explaining the world. And yet, as the one and the other, are in the same universe, and evolve at the same time, there must be a secret relationship.*"

The path towards the attractor requires diversity, different species, different cultures, ideas, ideologies and religions. Like the tiles of a mosaic which together form the unity of the design, our individualities are pieces which converging together give place to the design.

Steve Jobs found his mission in a computer that could be held in a hand, and this became his life project. Everyone has a purpose in life. Small or big they are all equally important. When we reach our goal we can die happily, and then continue the adventure towards the Omega point in a new life, with another mission.

- Life and death

Raymond Moody, an American psychologist and physician, became famous for his books on life after death and near death experiences, a term he coined in 1975 in his best-seller book *Life after Life*.

After a meeting with psychiatrist George Ritchie, who told him of an incident in which he died and had traveled in the afterlife, he began documenting reports of people who had experienced death.

Moody discovered that many elements are recurrent, such as the feeling of being out of one's body, the feeling of traveling through a tunnel, meeting dead relatives and of a bright light. After talking to over a thousand people who had this kind of experience, Moody started to support the idea that there is a life after death.

Moody noticed that people who die and are then resurrected thanks to modern medical techniques, come back deeply transformed. They often abandon their work to venture into activities aimed at the well-being of others. Moody underlines that near-death experiences are deeply transformative, they allow people to discover the meaning of their life and to connect to the great energy of love, what we here call the *Attractor*.

But do people have to experience death to begin this transformation process?

The answer was provided by Brian Weiss and Michael Newton.

As a psychotherapist and psychiatrist Brian Weiss was skeptical about reincarnation, but when one of his patients began to remember the traumas of a past life where he found the key to his recurring panic attacks and began channeling messages about Weiss's family and his dead son, Weiss began to use hypnosis to induce past life regressions.

Hypnotic trance is a state in which attention moves inward. We have continuous small hypnotic trances. Weiss found that a patient in a trance can easily live a previous life.

Michael Newton added hypnotic progression to hypnotic regression. After regressing his patients to a previous life, he used hypnotic progression to make them move to the point of death. This technique allows to experience death without having to die.

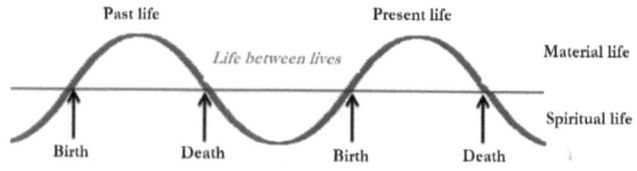

The idea is that we vibrate between life and death. When we are born syntropy is high, but the material world increases entropy and leads us to death. Death is the transition from the material to the spiritual life. In spiritual life syntropy increases to the point of having to be reborn. Spiritual life is syntropic and the connection with the Attractor is strong. Material life is entropic and the connection is more difficult: we do not remember what our mission and purpose of life are and with great ease we fall into the fascination of entropy and materiality. The goal is to reconnect people to the Attractor.

However, syntropy introduces a new concept of reincarnation that somehow contradicts or expands the model used by Weiss and Newton.

The unity of our soul is given by syntropy, by the fact that we converge towards the attractor. When we diverge the cohesive properties of syntropy diminish and our soul tends to shatter. This may explain numerous psychological and psychiatric disorders, such as the multiple personality disorder also known as dissociative identity disorder. This disorder is characterized by at least two distinct and relatively enduring personalities. Often there are problems in remembering certain events, beyond what would be explained by ordinary forgetfulness and these states alternate in a person's behavior.

Syntropy suggests that we reincarnate only if the syntropic (cohesive) component is strong, otherwise when we die our soul dissipates and loses its identity.

We can represent this as follows:

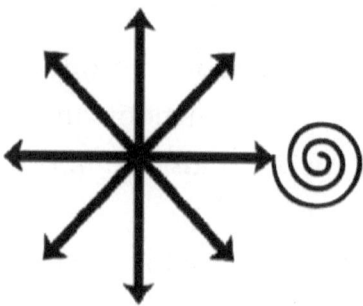

We are free to go in all possible directions, but only one converges towards the attractor and leads our soul to be cohesive, allowing to maintain its identity.

On the contrary, the identity of those who move away from the attractor vanishes with death. The identities of people who move partially towards the attractor will mix up leading to multiple experiences of past lives where we can be the reincarnation of a group of souls and not a single soul. According to Teilhard de Chardin the universe is gradually increasing its spirituality and eventually it will become a single soul that will unite with the Attractor in the Omega Point.

MIND AND CONSCIOUSNESS

Consciousness, the *"feeling of being alive"* is still a mystery. Neuroscientists assume that consciousness emerges from matter, whereas quantum scientists believe that matter emerges from consciousness.

Luigi Fantappiè and Pierre Teilhard de Chardin described consciousness as a property of the negative time energy. Physical energy can be perceived whereas the non-physical negative time energy can be felt: the head perceives, the heart feels.

We are constantly faced with what the head and the heart say, and we are forced to choose. The heart gives us direction and aim, whereas the head provides tools and experience. Both are needed.

Starting from the dual energy solution the mathematician Chris King speculates that free will arises from the fact that we are faced with bifurcations between information arriving from the past (*entropy*) and in-formation arriving from the future (*syntropy*).

These bifurcations entail choices and choosing puts us in a condition of free will.

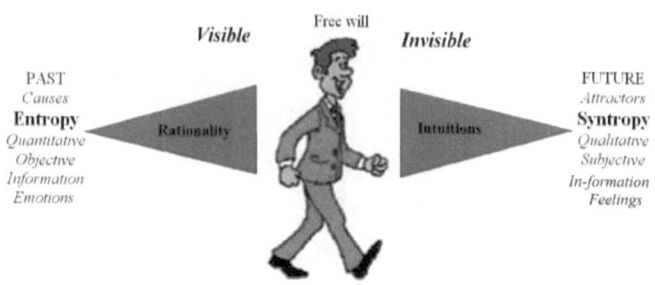

Supercausal model of free will

Since the forward and the backward in time energies are perfectly balanced, similar amounts of information and in-formation are received.

This might explain the perfect division of the brain into two hemispheres.

We can replace the previous illustration with that of the two hemispheres of the brain, where the left hemisphere is the seat of the "forward in time" logical reasoning and the right hemisphere is the seat of the "backward in time" intuitive reasoning.

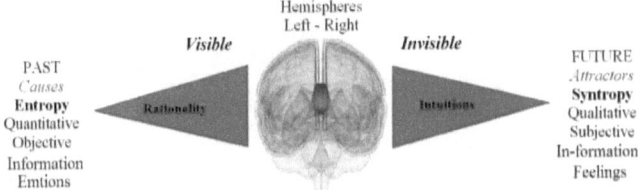

Where the rational-logical thinking is objective and quantitative ant the intuitive thinking is subjective and qualitative.

Syntropy adds to this picture the compass of the heart and the attractor and describes the mind as organized on three levels:

- the *conscious mind* which is associated to the head and free will;

- the *unconscious mind* which is associated to the autonomic nervous system and highly automated processes;

- the *super-conscious mind* which is the attractor, it is future oriented and provides direction, purpose and meaning to our life.

Conscious
Free will

Unconscious
Automated processes

Superconscious
Intuition
Finality / Meaning

The *conscious mind* on which we are tuned during the time we are awake, connects us to the physical reality. The conscious mind chooses between feelings that come from the autonomic nervous system, i.e. the unconscious mind, and information that comes from the physical plane of reality. This continuous state of choice is at the basis of free will.

The *unconscious mind* governs the vital functions of the body, therefore called involuntary, such as heartbeat, digestion, regenerative functions, growth and reproduction.

In addition, it implements highly automated programs, which allow us to perform many complex tasks, without having to think continuously about them, such as walking, riding a bicycle, driving, etc.

The autonomic nervous system supplies the body with syntropy and it is therefore the seat of feelings that inform us about the connection with the attractor. The unconscious mind can be accessed during dreams, or using techniques of relaxation and altered states of consciousness such as hypnotic trance.

The *superconscious mind* is our attractor, the source of syntropy, the energy of life, which guides towards wellbeing and happiness.

The superconscious mind provides us with a mission, a purpose, and uses intuitions, insights, dreams and visions. It provides intelligence, knowledge and answers to problems. It leads towards more intelligent and perfect designs which are the outcome of the contribution of all the individuals who share the same attractor.

- The conscious mind and free will

The conscious mind must constantly choose between future and past, and this process is at the basis of free will.

In-formation coming from the future acts as a pull factors, typically referred to as feelings of the heart, whereas *information* coming from the past acts as a push factors, typically based on memories, experiences, knowledge and acquired emotions.

We are constantly mediating between pull and push factors.

Past and future cohabit in our mind and require the specialization of the two cerebral hemispheres. The cortex is not a single block, but it is split in the left hemisphere which is the seat of linear thinking, based on causality, and the right hemisphere which has a global approach and is guided by feelings.

The left hemisphere sees the shape, how

things appear, whereas the right hemisphere feels the essence, the colors.

The left hemisphere is limited to exteriority, quantity and what is visible, the right hemisphere is focused on interiority, quality, feelings and what is invisible.

The neurophysiologist Antonio Damasio found that people with decision-making deficits, who are not capable of performing advantageous choices, show alterations in the ability to feel.

This deficit is common among people who have lesions in the frontal lobe of the brain or use drugs and alcohol which impair the ability to feel.

People with decision-making deficits have normal and intact cognitive functions: memory, attention, perception, language, abstract logic, arithmetic ability, intelligence, learning and knowledge. They respond normally to the majority of tests, and their cognitive functions are intact and normal, but they are not able to decide appropriately for anything that concerns their future. A dissociation is observed between the ability to decide on objects, space, numbers and words and the ability to decide advantageously for the future.

On the one hand, the cognitive functions are intact, but on the other hand these people are unable to use them advantageously. In neuropsychology this deficit is referred to as dissociation between cognitive abilities and their use.

Individuals with decision-making deficits are characterized by knowledge but not by feelings. They lack concern for the future, they are unable to plan for the future and make an effective program for the hours to come, they confuse priorities and lack insight and foresight.

Damasio shows that inner somatic sensations that take the form of feelings, acceleration of the heartbeat, hunches, contraction of breath, and muscles are fundamental in decision making.

In normal subjects, who decide advantageously, these inner feelings help to orient rationality, leading to an appropriate space in which the tools of logic can efficiently help the decision-making process.

Decision-making deficits suggest that there is a set of systems which orient thinking towards the future, towards an end, and that this set of systems is at the basis of decision-making and it is guided by feelings.

- *The unconscious mind and the autonomic nervous system*

The autonomic nervous system is in charge of acquiring syntropy and distribute it to the body, nourishing regenerative and healing processes, and providing the project, the shape, to the physical body and to its parts.

Attractors retroact from the future via the autonomic nervous system. At the same time attractors receives experiences from all the individuals linked to it and select what is advantageous to life, redistributing this knowledge to all the individuals as information.

According to this view evolution is a collective process which is guided by the intelligence arriving from the attractors.

The word intelligence comes from Latin and it is the combinations of two words: *intus*=inside and *legere*=read.

If we try to explain intelligence, order and in-formation as a result of past causes, we get into logical contradictions and paradoxes, since causality and random mutations are governed by entropy which leads to an increase in disorder. Nevertheless, we witness an incredible complexity and the convergence of this complexity towards common and intelligent designs, despite individual differences. Considering only the influx of the past it is impossible to explain why individuals converge towards the same designs, and the stability of these designs in time. Attractors which retroact from the future can instead explain this. Once a form takes place in the attractor, it can in-form all the individuals linked to it.

The autonomic nervous system plays a key role since it connects individuals to the attractor and in this way receives life energy, (ie syntropy) and in-formation.

Despite the incredible amount of intelligence that in-formation shows, it is widespread at all the levels of life. It is a property of the autonomic nervous system, ie the unconscious mind.

The autonomic nervous system:

— Is guided by feelings.
— It provides syntropy, vital energy, to the various organs of the body and performs healing actions based on the designs received from the attractor.
— It behaves like a mechanic who consults the book of the manufacturer to perform repairs and maintain the system as close as possible to the project. The project is not mechanical and instructions are written with the ink of e-motions.
— It underlies all the involuntary functions of the body and is responsible for controlling the motion of muscles and limbs.
— It governs all the functions of the body that are not subject to choice and which do not require the conscious level. For example, it is responsible for digestion, heart rate,

assimilation of food, cell regeneration. These are processes which are completely unknown to our conscious mind. We do not know how they are carried out and, often, we do not even know that they exist. The body knows everything and shows an extraordinary level of intelligence.

- It directs and regulates these processes, thereby expressing capabilities and intelligence which are incredibly higher than our conscious mind.

- It memorizes learning patterns of behavior which it then executes autonomously and automatically, and which are maintained over time, giving rise to habits and learning. This memory is then stored, at least in part, in the muscles of the body in the form of patterns of behavior.

- It repeats behavioral patterns, until they become habits that are activated automatically, regardless of our will. These patterns are then placed firmly in the memory of the unconscious mind. The conscious mind often does not remember what was included in the memory of the unconscious mind. Consequently accessing the unconscious mind can open incredible possibilities in the processes of knowing ourselves.

The unconscious mind also acts as a guardian of any information that the conscious mind cannot handle.

Nearly all visceral functions (heartbeat, breathing, digestion, etc.) are under the control of the autonomic nervous system which acquires syntropy. Since syntropy flows backward in time it activates visceral feelings in advance, providing information about the future. Visceral feelings alert about the future and animals follow them instinctively. This allows animals to feel the future with days in advance.

The first report dates back to 373 B.C., when animals, including rats, snakes and weasels, fled from the Greek city of Elice few days before a devastating earthquake. Animals panicked, dogs started barking and whining for no apparent reason.

In China, where the invisible energy of life

is taken seriously into account, these strange behaviors are used as alarm bells. For example, in 1975 people of Haicheng, a city with one million people, were ordered to flee their homes. A few days later a magnitude 7.3 earthquake destroyed the city. If the abnormal behavior of animals had not been taken seriously, more than 150,000 people would have died.

We often confuse feelings with emotions. Generally speaking we can say that feelings are linked to attractors and to the future, whereas emotions are linked to the past.

- The superconscious mind and the attractor

The superconscious mind is the attractor. It is outside our physical body and time and it is connected to our body via the autonomic nervous system (solar plexus/heart).

The attractor is the source of syntropy. Since syntropy acts as an energy concentrator, a good connection with the attractor is signaled by feelings of warmth and wellbeing in the heart area. In contrast, a weak connection with the attractor is signaled by feelings of void and pain usually named anxiety and anguish, accompanied by symptoms of the autonomic nervous system, such as nausea, dizziness and feelings of suffocation.

The superconscious mind provides aim and direction, intuitions and insights of the future.

The connection with the attractor is fostered when we reduce entropy in our life, when we calm the chatter of our mind, our fears, and avoid the use of alcohol, tobacco, drugs and coffee, when we nurture a good contact with nature, when we follow a vegetarian/liquidarian diet and a frugal and minimalist life style.

The invisible world of syntropy works in the opposite way to the ordinary one: richness requires frugality, unity needs diversity, strong effects want small actions.

Results otherwise impossible can be achieved with little effort, such as transforming deserts into fertile soil, reviving the process of rainfalls and reducing the greenhouse effect (see syntropic agriculture[39]); reduce debts and costs; meet the energy needs in an environmentally friendly and sustainable way; turn crises into opportunities, produce wealth and wellbeing.

The difficulty lies in understanding the language of the heart. When we use the compass of the heart we learn to choose in the most advantageous way for us and for the others.

To better understand how the superconscious mind works, it is worth quoting the words of the mathematician Henri Poincaré (1854-1912).

Poincaré noticed that when faced with a new mathematical problem he began using the rational approach of the conscious mind that allows to become aware of the elements of the problem. But, since the options are infinite and it would take infinite lives to evaluate them all, some other type of process leads to the correct option.

---

[39] https://lifeinsyntropy.org/en/

"The genesis of mathematical creation is a problem which should intensely interest the psychologist ... It is time to penetrate deeper and to see what goes on in the very soul of the mathematician. For this, I believe, I can do best by recalling memories of my own. ... all my efforts only served to show me the difficulty ... Thereupon I left for Mont-Valérien, where I was to go through my military service; so I was very differently occupied. One day, going along the street, the solution of the difficulty which had stopped me suddenly appeared to me. ... Most striking at first is this appearance of sudden illumination ... These sudden inspirations never happen except after some days of voluntary effort which has appeared absolutely fruitless ... I have spoken of the feeling of absolute certitude accompanying the inspiration ... the solution is felt rather than formulated ... It may be surprising to see sensibility ... the feeling of mathematical beauty, of the harmony of numbers and forms, of geometric elegance. This is a true aesthetic feeling that all real mathematicians know, and surely it belongs to sensibility."[40]

---

[40] Henri Poincaré, *Mathematical Creation, from Science et méthode*, 1908.

The process of creation can be divided into four phases:

1. A conscious phase during which we acquire the elements that make up the problem.
2. An unconscious phase that ends with the intuition, which is highlighted by a feeling of certainty and beauty.
3. Intuition is the starting point from which the conscious mind can formalize the details, thanks to the strict discipline and logical thinking of the conscious mind, of which the unconscious is incapable.
4. When the details are formalized the empirical validation ends the process.

When intuitions arise we experience a feeling of certainty, warmth and beauty that lets the solution arise to the conscious level of the mind.

The interaction between past and future, conscious and unconscious can be noticed in the strange strategy cats use when they want to jump on a table.

They are unable to see what is on the table, but they smell the food and want to get on it. They first start circling the table till they choose a spot. Then they start assessing the jump moving in a slow motion their back.

But what are they assessing, since it is impossible for them to see what is on the table? They cannot rely on any rational information for their assessment. And still, when they jump, they land perfectly in the most narrow spots!

According to syntropy they engage a game with their feelings, assessing in this way the future outcome. They try infinite invisible jumps and feel the results. When the feeling is of certainty and they jump. Feelings of certainty which accompany intuitions highlight the solution. Similar feelings are triggered by the attractor which provides us with life energy and purpose.

- *When does consciousness end?*

The concept of brain death was introduced in the scientific world at the same time to the first transplant of organs, since the criteria of natural death (end of heart activity and blood circulation) does not allow organ transplants.

Brain death is believed to cause the death of consciousness and of life.

This assumption is used to legitimate organ transplants from warm bodies.

The first definition of brain death was developed in 1968 by an ad hoc committee set up at Harvard Medical School.

The Harvard criteria for brain death determination have now become the bases for national laws. These criteria establish when it is permissible to consider the patient "legally" dead. The Harvard criteria are also the bases for the laws on organ transplantation, since organs need to be removed from the donor when the heart is still beating.

In 1975 the second international symposium on brain death was held in Havana (Cuba). The central moment for the diagnosis of death is the cessation of all brain functions. Only then it is totally useless to continue to provide assistance to the patient and to declare the state of death.

For cessation of all brain functions a EEG is defined as a "flat EEG" when the amplitude is not greater than 2 micro volts, corresponding to about 5% of normal activity.

The superconscious model of the mind considers consciousness placed outside our physical body, entering the body through the solar plexus and not the brain. It is therefore connected to the activity of the heart and not of the brain.

This assertion is supported by the fact that, when explanting organs from a person who is legally defined as dead, with a low EEG activity, this person starts defending and screams and must be tied to the operating table in order to proceed to the explant of organs.

Furthermore, the number of people diagnosed with brain death, who wake in full consciousness is simply amazing.

According to syntropy, when the heart stops and the connection between the body and the attractor ends, then death occurs. Syntropy stops flowing and all the organs and tissues die, making organ transplant impossible.

In 1985, with a statement of the Pontifical Academy of Sciences, the Vatican accepted the Harvard Report. Pope John Paul II talked in several occasions on the topic legitimating the removal of organs from warm bodies, despite the fact that they are still breathing and with their hearts beating.

On September 3, 2008, *"L'Osservatore Romano"*, the Vatican newspaper, dedicated the editorial to the fortieth anniversary of the Harvard Report. In this editorial Lucetta Scaraffia declared that brain death cannot be used to assert the end of a life and the definition of death should be reviewed in the name of new scientific assumptions.

A few days after the publication of Scaraffia's editorial a note from the Vatican Press Office stated *"an article does not change the doctrine: it is an editorial in L'Osservatore Romano, signed by a person who brings the authority of that person."*

The reactions of the medical / scientific world were immediate: *"The criteria for brain death is the only scientifically valid criteria in order to sanction the death of an individual."* Moreover, *"the worldwide scientific community approves the criteria established by the Harvard report and the criticism that comes from fringe minority, are based essentially on non-scientific considerations."* Finally, *"scientifically advanced countries have accepted as the norm all the criteria of brain death."*

However, the debate within the scientific Catholic world continues to grow. A whole chapter in a book edited by Paolo Becchi: *"Brain death and organ transplantation. A question of legal ethics,"* published by Morcelliana of Brescia illustrates the ambiguity of the Vatican and contains the statement of Hans Jonas's who argues that the new definition of death established by the Harvard report was not motivated by any real scientific discovery, but by the need for organs for transplantation.

In 1989, the Pontifical Academy of Sciences had addressed the question and Professor Josef Seifert, Dean of the International Philosophical Academy of Liechtenstein, was the only one to object to the definition of brain death.

But, when the Pontifical Academy of Sciences met again to discuss the issue, on 3-4 January 2005, the positions reversed. The participants, philosophers, jurists and neurologists from various countries, agreed in considering that brain death is not death of the human being and that the criterion of brain death is not scientifically credible and should therefore be abandoned.

For the Vatican officials who subscribed the Harvard report these results were unacceptable and Bishop Marcelo Sánchez Sorondo, chancellor of the Pontifical Academy of Sciences, ordered not to publish the proceedings of the meeting.

A number of speakers gave their papers to an outside publisher, Rubbettino, and a book was published with the Latin title *"Finis Vitae"*, edited by Professor Roberto de Mattei, deputy director of the Italian National Research Council. The book was published in two editions, in Italian and English and contained eighteen essays, half of whom have been written by scholars who did not attend the meeting of the Pontifical Academy of Sciences, but shared its views, among which that of Professor Becchi.

- *Consciousness in China*

In Chinese ideograms consciousness is described using two ideograms: the ideogram of the heart 心 (xin) and the ideogram of the head 头 (tou):

The heart is placed in the first position, thus telling that the essence of consciousness is the heart, whereas the head is placed in the second position, thus suggesting that it is a tool of consciousness.

It is also remarkable to note that in Chinese ideograms an "idea" is the combination of the heart on the left and the ideogram "to think" 想 on the right. The ideogram "think" contains the ideogram of the heart as a radical:

When we communicate our thoughts to someone we have at the left "message" 信 and at the right the heart. In other words, our thoughts are "messages from the heart":

For insights and intuitions on the left of the heart there is the ideogram warmth. Intuitions are described as feelings of "warmth in the heart":

Being diligent, attentive, devoted to a project is described as "eye of the heart":

When in the course of our business we are scrupulous we use the ideogram "a lot" associated with the heart:

When we become actors of our choices, of our free will, we use the ideogram "force" associated to the heart, "a strong heart":

However, when we are depressed we talk about "grey heart" a "heart with no color":

Finally, when we are able to solve a problem, we talk about a "peaceful heart":

Ideograms suggest that when it comes to consciousness, attention should shift from the head to the heart.

This same consideration can be found in many ancient civilizations.

In ancient Egypt the heart was considered to be the seat of consciousness, whereas the brain was considered unnecessary fat material.

In ancient Greek, Roman, Indian, Arab, and Jewish civilizations, the scientific, medical, philosophical and mystical systems considered the heart the seat of consciousness, whereas the brain was a tool, the servant of the heart.

THE UNITARY THEORY
AND
THE THEORY OF EVERYTHING

Luigi Fantappiè was born in Viterbo, Italy, on September 15, 1901.

He graduated from the most exclusive Italian university, the Scuola Normale Superiore di Pisa, at the age of 21. During the University years he became good friend with Enrico Fermi. and was very well known among physicists.

After the dissertation Fantappiè moved to Paris and then to Germany giving lectures.

When he came back to Italy he was assigned at the University of Rome where he became full professor at the age of 27.

In the years 1934-1939 he was sent to Brazil to start the faculty of mathematics in San Paolo.

In April 1951 Oppenheimer invited him to become a member of the exclusive Institute for Advanced Study in Princeton and work directly with Einstein.

Fantappiè died during the night between the 28th and 29th of July 1956.

This is how Fantappiè described his Unitary Theory in a letter to a friend:

"*It was in the days just before Christmas 1941, as a consequence of conversations with two colleagues, a physicist and a biologist, that I was suddenly projected in a new panorama, which radically changed the vision of science and of the Universe which I had inherited from my teachers, and which I had always considered the strong and certain ground on which to base my scientific investigations.*

Suddenly I saw the possibility of interpreting a wide range of solutions, the advanced potentials of the wave equation which can be considered the fundamental law of the Universe. These solutions had been always rejected as impossible, but suddenly they appeared possible, and they explained a new category of phenomena which I later named syntropic, totally different from the mechanical, physical and chemical laws, which obey only the principle of causation and the law of entropy.

Syntropic phenomena, which are represented by those strange solutions of the advanced potentials, obey two opposite principles of finality and differentiation and they are not causable in a laboratory.

Its finalistic properties justify the refusal among scientists, who accepted without any doubt the assumption that finalism is a metaphysical principle, outside Science and Nature. This assumption obstructed the way to a calm investigation of the real existence of this second type of phenomena; an investigation which I accepted to carry out, even though I felt as if I were falling in an abyss, with incredible consequences and conclusions.

It suddenly seemed as if the sky were falling apart, or at least the certainties on which mechanical science had based its assumptions. It appeared clear to me that these "syntropic", finalistic phenomena which lead to differentiation and could not be reproduced in a laboratory, were real, and existed in nature, as I could recognize them in the living systems.

The properties of syntropy opened consequences which were just incredible and which could deeply change the biological, medical, psychological, and social sciences."

This theory unifies the physical, chemical, biological and psychological phenomena, including those of consciousness, in the same rational frame. It also provides interpretations of the fundamental phenomena of quantum mechanics.

It might seem strange that a mathematician adventured himself in such a wide exploration in the fields of other sciences, without having a specific knowledge of them. This consideration stopped Fantappiè in letting his theory become public. But when he outlined its content to the colleague and friend Professor Azzi of the University of Perugia and received a strong and positive support, he felt he had to formulate it in a more detailed way and discuss it with colleagues of other disciplines.

Fantappiè presented his *Unitary Theory* on November 3, 1942, in Spain, at a conference at the *Consejo Nacional de Investigaciones Científicas*. He then was invited to Barcelona by the *Academy of Science*, where on December 1, 1942, he discussed the details of the Theory in a private meeting.

On the days that go from the 31st of May

to the 2nd of June 1943 he was invited by Professor Carlini to the Science and Philosophy conference which was held at the *Scuola Normale Superiore di Pisa*. In this occasion he presented his Unitary Theory among scientists of the most diverse orientations and was able to discuss it with many prestigious colleagues, among whom professors Severi, Rondoni, Carrelli, Puccianti, Persico, Guzzo, Abbagnano and Banfi and was given an entire afternoon for questions and answers. It was then that he decided to write *The Unitary Theory of the Physical and Biological World*.

In this chapter Luigi Fantappiè's Unitary Theory is presented using mainly an adaptation of his works which is available at: www.amazon.com/dp/1520237529.

As Fantappiè shows, the Unitary Theory:

- confirms the law of causality and the second principle of thermodynamics for all the phenomena which we call entropic. Causality, which was a conceptual category, becomes a law of the entropic phenomena, which has a precise and objective meaning.

— describes phenomena totally different from the entropic ones, which we can find in the mysterious properties of life. These phenomena are predicted and explained by the same equations which govern the entropic phenomena, but are essentially different and allow to see an immense panorama, which might be more vast, diversified and meaningful of the entropic phenomena.

— shows that the same wave equation which combines special relativity with quantum mechanics predicts syntropic and entropic phenomena. Syntropic phenomena are moved by attractors, finalities, whereas entropic phenomena are moved by causes.

Scientists had postulated that using the principle of causality all natural phenomena can be reproduced. The Unitary Theory shows that only the entropic phenomena can be caused and reproduced, whereas syntropic phenomena cannot be caused and reproduced, they can only be observed.

All the knowledge that has been developed in the last centuries using the experimental method, on which science is based, is limited to the entropic side of nature, whereas for the syntropic phenomena we need a new scientific methodology.

Syntropic phenomena can be influenced indirectly from specific entropic phenomena, but on the whole they constitute an extremely important part of the universe which is beyond our possibility of manipulation.

The entropic side of reality will inevitably fail to account for the totality, since the laws of nature are symmetric in regard to time and can be diverging-entropic and converging-syntropic, and this last type of phenomena are those which are at the essence of Fantappiè's discovery.

If we look at the present knowledge of the intimate structure of the Universe, we see that it can be summarized in three basic points:

- Dalton's atomic theory established in the XVIII century and later improved by Stanislao Cannizzaro, with the distinction of molecules and atoms, and then by Lorentz who formulated the particle theory of electromagnetism and Planck and Einstein with the quantum theory of energy. These results on the intimate atomic-particle nature of matter of the entire Universe is now considered acquired, since it has been tested and validated for more than two centuries.

- The wave nature of all the physical phenomena, when considered in their most profound essence, at the level of quantum mechanics. Studied by Heisenberg, Schrödinger, Dirac, and others, has given birth to modern nuclear physics. The wave nature of the physical phenomena can now be considered acquired thanks to the experimental validation of Davison and Germer with electron rays which shows diffraction and interference properties in particles. These properties are typical of waves.

— The validity of the Theory of Special Relativity, which has received corroboration at the atomic level, such as the explanation of the increase in mass, the inertia of the electron, and the increase in speed. This theory leads to a description based on four dimensions which unites space with time, reaching in this way a perfect symmetry among the spatial and time dimension. This representation is named chronotype.

How can these three fundamental elements be harmonized?

First of all the atomic-particle nature of matter and the wave manifestation seemed to conflict, since one is deterministic and the other probabilistic.

At the moment this conflict has been solved saying that it is impossible to predict in a deterministic way the behavior of particles since the prediction is attributed to waves which are probabilistic.

Waves offer a deterministic prediction only when we consider large numbers of particles.[41]

In Boltzmann and Poincaré theory the Universe was described as governed by strictly deterministic laws, both at the macro and at the micro level. Probability was used in a way which was considered only to be temporary, with the belief that the evolution of science would have replaced the mean values of probability with the exact values of the rigorous deterministic laws, which were believed to be at the foundation also of the microcosm.

Now, instead, the probabilistic laws of these phenomena are considered to be at the foundation of the Universe, whereas the deterministic laws, which are valid at the macro level, are considered to be only a consequence of the law of large numbers.

---

[41] Wave phenomena are represented by differential equations with second order derivatives of the hyperbolic type, whereas in order to describe the phenomena studied by classical mechanics and by optics equations with first order derivatives are used (Jacobi equations) or the equivalent ordinary differential equations (canonical mechanical equations). This implies that whereas in classical mechanics we can distinguish trajectories of entities with their own individuality, in wave mechanics the presence of equations with partial derivatives of an hyperbolic order greater than one leads to phenomena which are not localized, with the change of time, in a limited area (just think of the space occupied by a particle).

In 1927 Schrödinger renounced to special relativity in the formulation of his wave equation since in quantum mechanics waves should propagate at infinite speeds, and this is in conflict with the theory of special relativity which prohibits speeds greater than the speed of light.[42] The conflict between Schrödinger's non-relativistic wave equation and special relativity is obvious also at the general level,

---

[42] Schrödinger's wave equation takes the Hamiltonian function H, which characterizes the system in classical mechanics and measures the total energy relative to its space coordinates and to the momentums, and writes that the wave equation (which describes with the square of its modulus the probabilistic density) has a variation in time (a first derivative relative to time, using the mathematical language) which is proportional, for a constant factor, to an expression which is obtained applying to the same function a linear differential operator, which is obtained from the Hamiltonian function replacing the momentums with the derivatives of the corresponding variables, changed using a constant factor. Since the Hamiltonian function is squared for the momentums, a linear expression of the second derivatives is obtained referring only to the spatial variables, and a term which contains the unknown function y (which is relative to the potential), and a last term in which the first derivative is relative to time. In the case of a single particle with the space coordinates x, y, z, Schrödinger's wave equation is a linear differential equation of the second order, which contains the first derivative relative to time, and the second derivatives of the space variables are always parabolic (since the particle is a H term which is expressed by a polynomial of the second order in the momentums), of the same kind of the equation that governs the conduction of heat in solid matter.

since time appears in a non-symmetric way, as a first derivative.

It is generally accepted that Schrödinger's wave equation is only a temporary description of the quantum phenomena, which is valid with good approximation only in those cases in which the speed of light can be considered infinite, but which will have to be replaced by a quantum-wave theory which is more exact and agrees with special relativity.

On the contrary relativistic wave equations are symmetrical for all four variables, the space variables x, y, z and the time variable t, in agreement with special relativity. In this way a second order equation is obtained not only for the space variable, but also for time, and the D'Alembert operator is used.

The study of such an equation was brilliantly conducted by Dirac, considering all its implications, in the case of the electron, decomposing the equation of the second order in an equation of the first order, and showing that this wave-relativistic equation of the electron allows the full explanation of phenomena that until then where difficult to understand rationally, such as the magnetic momentum of the electron, which we now call the spin, which is due to the rotation of the electron on itself. Dirac found in his equation that beside the usual electron, also a symmetrical solution appeared, a neg-electron which is now named positron, which had not been observed since then and which was considered to be impossible.

But after a short time, the positron was discovered by Blackett and Occhialini, and this validated the prediction that Dirac's equation made of this particle, showing at the same time the strong foundation of quantum mechanics when combined with special relativity.[43]

---

[43] The most important properties of the second derivative equation which was initially formulated by Dirac are obtained from the characteristic cone, which is determined by the second order terms of the equation. These terms are found applying the D'Alembert operator to the unknown function, and consequently the characteristic cone is always real, matching the chronotype which, with the vertex in the assigned event, divides the events from the future to the past ones and from those which can be concomitant, according to Special Relativity. Consequently from this structure of the characteristic cone the value of the unknown function y of the assigned event (that is to say in the point of the chronotype with coordinates x,y,z,t), at least in the case of the events which we have previously determined, can depend only on the

It is important to underline that although we don't yet have the details of the partial derivatives equations which describe in all their details the various quantum systems, we can determine some very important characteristics of these unknown differential equations, such as the fact that the properties of the characteristic cone will apply to all, and the fields of dependency and influence of the solutions, which are described by Dirac's equation.

These properties have been deduced from

---

values of y and eventually on the terms of the equation (which represents the density of the distribution of the sources of the wave propagation) known from the past events, whereas the value of the y point and of the known term can influence only the values that y acquires in the field of the future events. In other words the field dependence of the solutions of the event which has been considered is attributed only to the past events, whereas the field influence to the future events, whereas events outside of the chronotype cannot influence or be influenced by the event. For those who are less familiar with the four dimensional representation of the chronotype, it is sufficient to say that the past events, that is events which fall within the boundaries of the cone, are given for each instant before the one we are considering t, by the points within a sphere with its center in the points x,y,z with a radius which decreases with the speed of light, till it reaches zero in the instant t, whereas the future events are given, for each instant following t by the points of a sphere, with the same center, with a radius which increases with the speed of light, starting from the zero value at the instant t.

those of the D'Alembert operator, which is linked only to the geometrical nature of the chronotype, and does not depend from the particular properties of the particle, which are instead described by the other terms of the equation which do not influence at all the geometrical nature of the chronotype. The chronotype does not vary when we consider a different type of particle, or particle systems, we will have that also for the equations of unknown partial derivatives, which support these quantum systems, the characteristic cone and the fields of dependency and influence of the solutions will be the same of those that Dirac found in his equations.[44]

---

[44] This can be clearly stated following another path; if we just consider that in wave phenomena the partial derivatives equations which describe them need to be of the hyperbolic type, and need to satisfy special relativity, the values of the solutions of a point x,y,z at an instant t, for any phenomena which we have caused, must be the consequence of values within the converging sphere towards the point at the speed of light (past events according to special relativity) and can effect only those points within the sphere which diverges from the same point, with the same speed (future events according to special relativity), otherwise if an element outside these two regions could affect or be affected from the event, the action between the two events should propagate at speeds which are greater than the speed of light, which according to special relativity is impossible.

The fundamental solutions of the D'Alembert operator have been provided by Poincaré[45], Ritz[46] and Giorgi[47]. A first solution describes waves diverging from the source and are named *delayed potentials*.[48] A second solution describes waves converging to the source and are named *advanced potentials*.

The criticisms to the possibility of advanced waves were made mainly by Wiechert, Lorenz, Poincaré, Ritz and Giorgi, who considered that if converging waves existed it would be possible to concentrate energy and in this way to devise a perpetual motion machine. And this was considered to

---

[45] H. Poincaré, Electricité et optiqtee, 2.e éd., Paris, 1901

[46] W. Ritz, Recherches critigues sur l'électrodinantique générale, Ann de physique, 8 s., t. 13, 1908, p. 145

[47] G. Giorgi, *Sulla sufficienza delle equazioni differenziali della fisica matentatica*, Rend. Lincei, s. Ga, vol. VIII, 1928. Per un'ampia bibliografia sull'argomento, cfr. A. Cabras, Sulla teoria balistica della luce, Mem. Lincei, s. 6a, vol. III, f. 6°, 1929.

[48] Starting from the hypothesis that the wave always starts from a source, with a density measured by the second known member of the equation; this solution is obtained in each point as the sum (integral) of the infinitesimal contributes (potentials) due to the sources, distributed in the single elements of the volume, in previous instants (to that which is being considered) at a certain time, that is needed for the wave to diverge at the speed of light c, from the volume element where the source is situated at the point considered;

be impossible.

Now, let us see how the notion of cause and causality, as they are understood by physicists and modern scientists, differ from the more general "deterministic principle", considered as the possibility of making a prediction.

When we say that the event A causes B, we believe that once we have observed A we can certainly predict that B will become true. But, we can also predict that after the event of night the Sun will rise, however no one can say that the rise of the Sun is caused by the night. In the notion of causality there is something more.

When can we say that A causes B?

The answer to this question must be searched in the experimental method, which Galileo put at the foundation of all the modern sciences.[49]

A is the cause of B when we insert experimentally A and we observe B.

_____

[49] The definition of cause which we give here coincides with the definition that Galileo gave: "*A cause is that which when present is followed by an effect and when removed the effect disappears.*"

But in order to have a convincing experiment we need to be free, at least within certain boundaries to cause A, where and when we wish. As a matter of fact if someone would want to convince that A is the cause of B producing A in order to asses B, only in a specific place and time, we would remain skeptic.

The experimental method provides an exhaustive answer to the question if A is the cause of B, only when we have the total freedom to produce A and see if B follows. Only in this condition we can be sure that A is the cause of B. This leads to the important conclusion that we can recognize the events which are the cause of others only thanks to the free will of the experimenter.

Causality gives way to the more general and objective "determinism" which tries to determine past and future events analyzing present events. But also determinism has shown to be insufficient in the study of particles, leaving the field to a wider perspective in the microcosm, which is based on probability.

We can state that widening our knowledge the categories which we were trying to apply have widened, moving from the law of causality, to determinism, to the modern probabilistic theories of quantum mechanics.

This does not mean that causality and determinism should be abandoned, but they cannot be used to explain all the reality.

Causality and determinism are certainly useful and fundamental in the study of a well-defined parts of reality. When we move from wave mechanics to the more limited deterministic field of the macrocosm, where the law of large numbers applies, probabilities change into frequencies which can be handled in a deterministic way.

If we isolate the system in such a way that nothing happens beside what the experimenter wants with his free-will and B is different from zero only from the moment when A is produced, we can state that A causes B. The cause becomes the source which causes B and, therefore, each event B which is caused by A, is always affected by diverging waves from the point A. The solution that governs B will therefore be of the type of the *delayed potentials*.

This implies that causable phenomena are always entropic. Each entropic phenomenon, each phenomenon based on diverging waves has its cause in the source from which the diverging waves originate.

In this way we get to the fundamental theorem:

A necessary and sufficient condition for B to be entropic, is that it can be caused using another phenomenon A, which is the source from which the diverging waves that constitute B are emitted.

The majority of the physical and chemical phenomena, which we can study in our laboratories, are entropic.

Causality applies to entropic phenomena, such as those studied in mechanics, acoustics, optics, electromagnetism and chemistry. This does not exclude that in nature we can have other phenomena, beside the entropic ones, such as the syntropic phenomena, which cannot be caused using our free-will, since they would then fall within the entropic phenomena.

Diverging waves imply necessarily the second law of thermodynamics, which states that entropy does not diminish, but increases during time.

From an intuitive point of view we can consider entropy as a state of leveling of a large number of particles. Diverging waves dilute in spaces which are always bigger, and if the space is limited, as it happens in a container, their intensity tends to level.

The wave equation extends this law to all the phenomena which are governed by diverging waves and in this way the second law of thermodynamics is no longer obtained from a probabilistic postulate, such as Clausius' principle of the elementary disorder, but it is a logical and necessary consequence of the law of causality. When the law of causality applies to a phenomenon, we can say that this phenomenon is entropic.

This is the reason why it is impossible to obtain a perpetual motion machine. The degradation of energy is a necessary and logical consequence of the law of entropy which applies to all the machines. The main argumentation which is used in order to exclude advanced potentials is that they would allow to devise perpetual motion machines, converging the energy that was first dispersed towards a point and then diverging it, then again converging it, and so on forever.

The main characteristics and properties of those phenomena which are constituted by advanced waves, which Fantappiè named syntropic, are profoundly different from the entropic phenomena previously described:

— They *cannot be caused* by our free will, at least in their essential components constituted by the converging waves, since on the contrary they would fall in the category of the entropic phenomena, which are governed by the law of causality, and characterized by diverging waves. For the same reason, syntropic phenomena can be influenced, in their evolution, only indirectly by specific entropic phenomena, the only which we can use, which can interfere with, for example by modifying the environment in which they take place, since it is plausible that if the two phenomena exist they are not separated in nature, but intertwined.

— They *concentrate energy* within always smaller spaces. Also the particles represented by these waves progressively concentrate in the center of the waves. Whereas the entropic systems go from concentrated to dispersed, in the syntropic phenomena exactly the opposite happens. We first have dispersed phenomena which concentrate in always smaller spaces. The entropic phenomena manifest with dissipative characteristics. An example is when we light a match. We have a cause which is concentrated in a small space, from which the light irradiates, with an intensity that diminishes with the distance, diluting the effect. Syntropic phenomena manifest with an anti-dispersive character, a converging manifestation, which goes from diluted to concentrated in specific points. Whereas the entropic phenomena radiate from specific points, syntropic phenomena concentrate towards specific points.

— The *concentration of energy cannot be endless.* Since it cannot continue indefinitely, after a period of syntropic concentration entropic dissipation takes over. This means that we witness a process of exchange of matter and energy. Incoming energy and matter indicate syntropic processes, outgoing energy and matter indicate compensatory entropic processes.

— *Entropy diminishes,* since with time differentiation increases. From a rigorous formal point of view syntropy has the same value of the second law of thermodynamics.

— We see a *tendency towards differentiation and complexity.* Syntropic phenomena show in complex forms, as it happens with biological systems which cannot be explained in a satisfactory way by using only their physical and chemical properties.

— They are *in a continuous state of energy*

dissipation (warm bodies), and this is a consequence of the fact that syntropic systems absorb energy but they don't evolve towards heat death.

It is possible to scientifically study syntropic phenomena considering that the D'Alembert equation is time reversal. This equation is symmetrical in respect to time.

Reversing the time variable all the solutions of the delayed potentials become solutions of the advanced potential, and vice versa. Consequently, a very simple way to obtain the syntropic properties of a system from the entropic ones is just to invert the time direction.

Nearly all the phenomena are dual phenomena. In our language this is usually expressed by adding the prefix "anti": combustion becomes anti-combustion, filtration anti-filtration, matter anti-matter, energy anti-energy, etc... Applying this principle of duality we can obtain the characteristics of the syntropic phenomena from its dual entropic phenomena.

According to the D'Alembert equation, entropic phenomena are activated when waves start diverging from the source. For example when we light a match electromagnetic waves start diverging at the speed of light in all the directions in a uniform way.

When we reverse the flow of time the dual syntropic phenomena shows. Waves concentrate towards the center of the sphere, increasing their intensity. These waves would be uniformly distributed in all the directions, independently from where they seem to come.

Let us consider the waves which propagate on a pond. We can cause this phenomenon, which is therefore entropic, by throwing a stone in the pond and observe how the waves propagate and diverge. The dual syntropic phenomenon would show these waves perturbations concentrate in a point from which the stone would then emerge, leaving behind the water at rest. If we could observe such a phenomenon we would think that some sort of intelligent being had organized it.

Now, let us imagine a brand new telescope

that we have forgotten in our garden. At first rust forms, then it falls and breaks into pieces. Pieces of metal and glass gradually deteriorate and mix with the ground. Changing the time flow we would see that from the ground different pieces of metal and glass separate, then they find their place in a design of lenses and tubes which form the telescope until a brand new and perfectly functioning telescope is reached.

What puzzles us is the finalistic aim, which we usually attribute to the action of an intelligent being. Syntropic processes express finality, a purpose, intelligence as if a will is acting on them.

Finality is the characteristic of the syntropic phenomenon.

The law of causality and the law of finality are logical consequences of the intimate duality of the fundamental laws of physics. It is possible to state that without causes entropic phenomena cannot exist and without finalities syntropic phenomena cannot exist. Without causes and finalities the wave equations would be null. Consequently finality is not an accidental manifestation in a syntropic phenomenon, but a necessary condition of the syntropic phenomenon, without which it could not exist.

Science has investigated the entropic physical and chemical characteristics of life, without grabbing the essence of life. It is now well acquired in biology, thanks to the experiments devised by Pasteur, that there is no possibility of spontaneously producing life without starting from a minimum amount of life. This is referred to using the Latin words «*vivum nisi ex vivo*». Life stems from life. It is impossible to create life at our will. The non-causability of life tells that it is a syntropic phenomenon. It is also well known that vital phenomena cannot be influenced directly, but only indirectly. For example we cannot produce directly a plant or an animal with our hands, but we can only grow or raise them.

All living organisms concentrate in their body matter and energy. This tendency is visible especially in plants and it is due to the chlorophyllian process.

We can therefore assume that in plants

there is a quantitative prevalence of the converging syntropic phenomenon, which is also present in animals in their growth stage and then it is balanced with entropic processes at the adult stage, which start becoming gradually more relevant with aging and then totally prevailing with death.

It is interesting to note that in metabolism the syntropic processes of absorption of matter and energy and construction of structures are named *anabolic*, whereas the entropic processes of dissipation, destruction of structure and release of energy and matter are named *catabolic*.

The syntropic process of energy absorption is always coupled with its dual phenomenon of energy dissipation. One of the major properties of life is that it is constantly releasing energy. This constant release of energy and by-products is coupled with the assimilation of matter and energy. A process of exchange of matter and energy which is named metabolism.

During the growth period, anabolic processes are prevalent and an increase in differentiation is observed.

It is interesting to note that the probability that the smallest protein molecule arises by chance is less than 10^{-600}. This is an incredibly small number, represented by a 0 followed by 600 zeros and at the end, on the right, the number 1. In other words, the spontaneous formation of the smallest life molecule results to be practically impossible. The incredible amount of proteins that life shows conflicts with the second law of thermodynamics. This means that the law of entropy does not apply to life and that life is not an entropic phenomenon.

Finality is the fundamental characteristic of any syntropic phenomena, similarly to the principle of causality which is the fundamental characteristic of any entropic phenomena.

Only thanks to the principle of finality we can logically understand the smallest and most complex architecture of the living systems. Organisms differentiate in organs which are harmonically coordinated and arranged in order to reach a purpose. For example, the development of the eye starts from cells which are very similar, which then differentiate and take place in such ways that they build the elements of a perfect eye, such as lenses, vitreous body, which are by far more complex of a single protein.

The principle of finality shows that pretending to understand life through its physical and chemical elements, which are governed by causality, is just an illusion. Finality on which life is founded is similar and dual to the principle of causality which governs the entropic systems. Causality is the essence of the physical world, finality is the essence of life. Living systems tend towards aims and purposes. Life systems have a mission, and the greater the mission is, the more complex is the living system, with complex organs meant to reach its purpose.

The difficulty with the principle of finality is commonly found in the various theories of evolution. If we examine the most popular one, Darwin's theory of evolution, we see that it is based on three facts: the variability of life forms, the fight for survival, and the long permanence of life on Earth. These facts cannot be denied, but are not sufficient to explain life and all the various species of organisms.

In 1865 Mendel's experiments on plant hybridization seemed to prove the theory of evolution which Charles Darwin had published in 1859. But with Mendel we are not witnessing the formation of new species, we are witnessing the separation of genetic information into different characters and forms.

According to Darwin at the beginning on Earth only few simple unicellular life systems could exist.

Darwin introduces the concept of random variability as the cause of new species. About randomness, the probability of the random formation of any living system can be calculated using the kinetic theory of gasses which considers all the possible combinations with the same probability. Using this assumption the probability of the formation of the smallest protein is less than 10^{-600}. It is therefore easy to imagine how smaller the probability of the formation of an organ is, such as the eye, the ear, or any of the apparatuses that we commonly use. The probability of the formation of a whole animal is even smaller. The random permutations which are required for the formation of just one protein are greater than all the possible permutations in the history of the entire Universe. Consequently, the long permanence of life on Earth is insufficient to account for the formation of the smallest forms of life and of any living being. The probability of life happening by chance are by far smaller than the probability of witnessing water freezing when put in a pot placed on the flame of a cooker.

And, if life is caused it should obey the law of entropy and go towards the dissolution of any form of organization and complexity. With time we would see the increase of entropy and it is illogical to pretend that complexity can be achieved at the expenses of other beings or using the light of the Sun since in the first stages of the evolution of life on Earth, there weren't other beings and the atmosphere did not allow Sun rays to reach the land.

When on the contrary we consider life as a syntropic phenomenon, the principle of finality applies and leads to increase differentiation, complexity and harmony.

The planet Earth can be considered as an immense living organism. The fact that species are interdependent, that they cannot live without others, for example fruits need insects for the pollination, we need vegetables … all these species can be considered as parts of a more complex organism orchestrated by a finality, which can be reached only through differentiation.

In human beings cells cooperate towards greater ends and only in pathological situations, when they lose their end, they develop in an excessive way, suffocating other cells, as it happens with cancer.

At the beginning of evolution simple forms of life are the aim, then they become the foundation blocks for always higher forms of life. Species are not caused by previous species, but they are attracted towards future designs and forms.

Syntropy solves the profound dissymmetry that the second law of thermodynamics has introduced in the universe, by considering all the solutions of the fundamental equations. The theory of syntropy shows that the solutions that physicists wanted to exclude represent exactly the essence of life phenomena, that seemed impossible to be explained.

Syntropy is capable of unifying different scientific disciplines in a harmonic way, opening in this way the road to a unified theory, a theory of everything that encompasses in a coherent theoretical framework all the manifestation of the universe.

With the formulation of the experimental method the problem of science was considered definitely solved. This method considers causality at the foundation of all natural phenomena.

The experimental method is used to test cause and effect relations. In the case of positive results the hypothesis is accepted, otherwise it is rejected. Experiments provide the verdict which allows to separate what is true from what is false.

The experimental method is profoundly different from the method which Aristotle suggested, which was useful in the formulation of theories but did not provide a way to choose among the various hypotheses.

The experimental method implies the law of causality and has limited scientific investigation to entropic phenomena. We can therefore call the Galilean science an entropic science.

The experimental method is divided in three steps: observation, formulation of a theory, experimental validations of its hypotheses.

As we have previously seen each entropic phenomenon has a dual syntropic phenomenon and vice versa. Consequently, although it is impossible to use the experimental method to test directly a syntropic hypothesis, we can set up an experiment in order to test the dual entropic hypothesis. In this way the study of the syntropic phenomena can be done indirectly studying the dual entropic phenomena.

Syntropic scientists would therefore have to search for the dual entropic phenomena, since when they manage to do this it is possible to progress using the experimental method.

Let us apply this dual method to a phenomenon which has yet to be explained, such as the absorption of water and nutrients from the land and their rise in the higher parts of the plant.

The hypothesis of osmosis does not stand since plants also acquire salts from the land. The idea that capillary conducts are responsible for the rise of water also does not stand when we consider that some trees can reach the height of 150 meters. These phenomena of absorption of water and rise of water seem to contradict the entropic laws of physics and this suggests that we are in front of syntropic phenomena which cannot be caused artificially. We can therefore apply to them the method of "dual experimentation".

In order to obtain the dual entropic phenomenon let us imagine that time flows in the opposite direction. We would see the lymph flow down until it reaches the roots and then water and salts disperse in the land. This dual image can be reproduced, for example, putting a non-living pole in the land and observing how water and salts filtrate from the top to the bottom and through the land. This entropic process of filtration, which can be easily caused in any moment proves that the process which we are witnessing in plants is the dual process of filtration. We can therefore name it anti-filtration.

One may object that in filtration gravity helps the process. Well, when we change the direction of time also gravity changes and from an attractive force it becomes a diverging repulsive force which helps water rise in the anti-filtration process which we observe in plants.

Now, let us take the combustion of vegetal tissues. This is a phenomenon which we can cause at our will and which is therefore certainly entropic. We see at the beginning a highly differentiated body, which is made of complicated carbon structures which absorbs oxygen from the air and when burned emits carbon dioxide, water, heat and produces a red light.

When the time process is reversed shifting from entropic to syntropic we would expect carbon dioxide, water, heat and red light frequencies to be absorbed. This would leave the complementary radiation to red which is green. If we look around we will notice that this syntropic process of green color really exists. This is the chlorophyll process, in the green leaves of plants which absorb carbon dioxide, water and heat. The chlorophyll process is therefore the dual process to the entropic one of combustion.

Studying and determining the laws of combustion in our laboratories can therefore allow us to account for the dual property of chlorophyll.

It is interesting to note that consciousness, the will and human personality, are processes which are oriented towards the future, moved by finalities and not causes. We can therefore state that psychical phenomena, our will and personality can generally be considered syntropic phenomena. For this reason they cannot be studied exhaustively using the experimental approach. It is also interesting to note that actions such as impulsive and emotional reactions which are caused by something that happened in the past are also those in which the activity of consciousness is reduced.

What makes life different is the presence of syntropic qualities: finalities, goals, and attractors. Now as we consider causality the essence of the entropic world, it is natural to consider finality the essence of the syntropic world. It is therefore possible to say that the essence of life is the final causes, the attractors. Living means tending to attractors.

The law of life is not the law of mechanical causes; this is the law of non-life, the law of death, the law of entropy; the law which dominates life is the law of finalities, the law of syntropy. But how are these attractors experienced in human life? When a man is attracted by money we say he loves money. The attraction towards a goal is felt as love.

This suggests that the fundamental essence

of life is love:

"*I am not trying to be sentimental; I am just describing results which have been logically deducted from premises which are sure. The law of life is not the law of hate, the law of force, or the law of mechanical causes; this is the law of non-life, the law of death, the law of entropy.*"

The law which dominates life is the law of cooperation towards goals which are always higher, and this is true also for the lowest forms of life.

In humans this law takes the form of love, since for humans living means loving, and it is important to note that these scientific results can have great consequences at all levels, particularly on the social level, which is now so confused.

"*The law of life is therefore the law of love and differentiation. It does not move towards leveling and conforming, but towards higher forms of differentiation. Each living being, whether modest or famous, has its mission, its finalities, which, in the general economy of*"

the universe, are important, great and beautiful…
Today we see printed in the great book of nature - that
Galileo said, is written in mathematical characters -
the same law of love that is found in the sacred texts of
the major religions."

SYNTROPIC METHODOLOGY

Science (from Latin *scientia*, meaning knowledge) is a systematic enterprise that builds and organizes knowledge in the form of testable explanations and predictions. An explanation is a set of statements which clarify the relations among causes, context, and consequences of facts. Explanations may establish rules or laws which allow to formulate predictions. Consequently, relations (among causes, context and consequences) are at the basis of explanations and predictions and, when relations are studied in a replicable and objective way, it is possible to talk about science.

In the last four centuries science has been using the experimental method, however syntropic methodology requires a different method which is generally known as the methodology of concomitant variations.

Let's start describing the experimental method.

The experimental method is based on the *methodology of differences*, which John Stuart Mill described in the following way:

> "*If an instance in which the phenomenon under investigation occurs, and an instance in which it does not occur, have every circumstance in common save one, that one occurring only in the former; the circumstance in which alone the two instances differ, is the effect, or the cause, or an indispensable part of the cause, of the phenomenon.*"[50]

The methodology of differences works as follows:

- two similar groups are formed (they are named the experimental and the control group).
- Treatment (the cause) is given only to the experimental group and all the other conditions are kept equal, so that the control group differs from the

[50] Mill J.S. (1843), *A System of Logic*, University of Toronto Press, 1843.

experimental group only for the treatment.

- Consequently, any difference observed between the experimental group and the control group can be attributed solely to the treatment, because only this condition changes between the two groups.

In order to have similar groups, randomization is used in the belief that it should distribute evenly all the intervening variables, between the experimental and the control group. But, generally speaking, no controls are performed in order to verify if the condition of similarity is satisfied and often the experimental and control groups are different ever since the beginning of the experiment. A single subject with extreme values can produce differences which are not due to the cause (ie treatment), but are due to the initial dissimilarity of the control and experimental groups.

In order to test the effect of a drug the experimental procedure is the following:

- two similar groups are formed, assigning subjects randomly to the experimental group or to the control group.
- The drug is given only to the experimental group, while all the other circumstances are left similar. The control group is therefore given a placebo, a similar substance which has no effect.
- The differences observed between the two groups can be attributed solely to the effect of the drug.

Differences are the effect and the drug (also called treatment) is the cause. The following conditions are required:

- In order to study differences between groups it is necessary that the effect can be _added_ among the experimental subjects. For example, if a drug increases in some subjects the reaction times, whereas in others it reduces the reaction times, when adding these opposite effects a null effect is obtained. The effect exists, but it is invisible to the experimental methodology based on the study of differences.

– Differences can be calculated only when using *quantitative data* (ie data which can be added together). On the contrary, qualitative data cannot be added and it is unsuitable when using the experimental method.

– All possible *sources of variability must be controlled*. It is important that nothing, besides the treatment (ie the cause), can influence the variability of groups. For this reason a controlled environment, which allows to keep alike all the possible sources of variability and in which each subject is treated exactly in the same way, is needed. Controlled environments require laboratory settings, which are very different from the natural context. The need for controlled settings limits the experimental method to analytical knowledge, detached from the context and from complexity.

– It is possible to study differences considering only one cause at a time or at the most few causes when studying their interaction.

- When samples are small (less than 300 subjects), randomization does not guarantee the similarity of groups, and differences between groups may not depend on the treatment, but on the initial diversity of groups.

Common mistakes:

- Differences can be caused by single extreme values. Just one single outlier[51] can cause statistical significant results and lead to assert effects that do not exist. Outliers are often kept or removed in order to manipulate results.
- In statistics, data transformation refers to the application of a deterministic mathematical function to each point in a data set which is replaced with the transformed value. A common example are logarithmic transformations. In

---

[51] In statistics, an outlier is an observation that is distant from other observations.

theory, any mathematical function can be used to transform the data set. Operating in this way, it is often possible to obtain differences between the two data sets, when there are no effects.

– When the effect shows in opposite directions, differences cannot be assessed and the effect becomes invisible.

From a statistical point of view the methodology of differences uses parametric statistical techniques which compare mean and variance values, such as Student's t and the analysis of variance (ANOVA). These techniques require that effects can be added, that data is quantitative and normally distributed (according to a Gaussian distribution), and groups are initially similar and are from the same population. But, these requirements cannot be met in life sciences and parametric techniques end producing results that are inconsistent. It is therefore of no surprise that a study published on JAMA (Journal of the American Medical Association), which revisited the results produced using the experimental method (ANOVA) and published in the period from 1990 to 2003 in 3 major scientific journals and cited at least 1,000 times, found that a study out of three was refuted by other experimental works. This finding raises serious doubts about the experimental method, when used in life sciences.[52]

In May 2011 Arrosmith published in the Journal Nature a study which shows that the ability to reproduce the results from phase 1 to phase 2 decreased in the period 2008-2010 from 28% to 18%, despite results were statistically robust in phase 1 (phase 1 indicates studies conducted on small groups, generally not exceeding 100 subjects, whereas phase 2 indicates studies conducted on larger groups, usually not exceeding 300 subjects).[53]

Gautam Naik in the article "*Scientists' Elusive Goal: Reproducing Study Results*" published on the Wall Street Journal on December 2, 2011

---

[52] Ioannidis J.P.A. (2005), *Contradicted and Initially Stronger Effects in Highly Cited Clinical Research*, JAMA 2005; 294: 218-228.
[53] Arrosmith J. (2011), *Trial watch: Phase II failures: 2008-2010*, Nature, May 2011, 328-329.

points out that one of the secrets of medical research is that the majority of results, including those published in major scientific journals, cannot be reproduced.

Reproducibility is at the foundations of making science and when results are not reproduced the consequences can be devastating.[54] Naik notes that researchers, particularly in universities, need to find positive results in order to publish and receive funding.

In the December 23, 2010 article entitled *"The Truth Wears Off,"* published in The New Yorker, Jonah Lehrer quotes a passage of a letter from a university professor, now an employee of a biotechnology industry:

> *"When I worked in a university lab, we'd find all sorts of ways to get a significant result. We'd adjust the sample size after the fact, perhaps because some of the mice were outliers or maybe they were handled incorrectly, etc. This wasn't considered misconduct. It was just the way things*

---

[54] Only in the US the biomedical industry invests each year more than 100 billion dollars in research

were done. Of course, once these animals were thrown out [of the data] the effect of the intervention was publishable."

There is plenty of evidence that the massive financial incentives lead to the suppression of negative results and the misinterpretation of positive ones. This helps explain, at least in part, why such a large percentage of randomized clinical trials cannot be replicated."

- The methodology of concomitant variations

In 1992 physicists at LEP (Large Electron-Positron Collider in operation at CERN in Geneva) could not explain some annoying fluctuations in the beams of electrons and positrons. Although very small, these fluctuations created serious problems when the energy of the rays must be measured with great precision. The experimental method did not provide any clue and in order to solve the dilemma the methodology of concomitant variations was used in order to test different hypotheses. Results showed the concomitant fluctuation in the energy of the particle beams of LEP and the tidal force exerted by the Moon. A more detailed analysis showed that the gravitational attraction of the Moon distorts very slightly the vast stretch of land where the circular tunnel of LEP is recessed. This tiny change in the size of the accelerator caused fluctuations of about 10 million electron volts in the energy rays.

The methodology of concomitant variations uses double entry tables of dichotomous variables. For example:

	Males	Females	Total
No accidents	50	105	155
Accidents	200	45	245
Total	250	150	400

Concomitances between sex and car accidents
(data invented for this example)

In this table the concomitance of the variable sex and car accidents is difficult to assess, since the total values of each column differs. When the absolute frequency values are converted into column percentage values it becomes easy to compare the columns "Males" and "Females":

	Males	Females	Total
No accident	50	105	155
	20%	70%	39%
Accidents	200	45	245
	80%	30%	61%
Total	250	150	400
	100%	100%	100%

Concomitances between sex and car accidents
(columns percentages)

We now see a strong concomitance between "*Males*" and "*Accidents*" (80%) and between "*Females*" and "*No accidents*" (70%). Concomitances are assessed according to the differences between observed frequencies (column percentage) and expected frequencies (percentages in the total column). For example, the expected percentage for "*no accidents*" is 39%, whereas in the "*females*" column we have 70%.

Since being male is determined before accidents take place, we can fall in the error of stating that being male is the cause of car accidents. However, this methodology allows to study intervening variables by splitting the table in two. For example, we can split the previous table in two groups: those who drive little and those who drive a lot:

	Drive little		Drive a lot	
	Males	Females	Males	Females
No accidents	70%	70%	20%	20%
Accidents	30%	30%	80%	80%
Total	100%	100%	100%	100%

Concomitances between sex, km driven and car accidents

In this table the concomitances between sex and accidents disappears. The correlation "*males-accidents*" is therefore mediated by the variable "*number of kilometers driven*", which is therefore an intervening variable. Consequently the relation becomes "*males drive a lot and consequently are involved in more accidents.*" Crossing three variable at a time allows to identify intervening variables and to study the context within which relations are valid. For example, when a concomitance is found between a drug and healing it is possible to study if it is true always, or only at certain conditions, such as specific age groups, sex, habits and other conditions.

The advantages of the methodology of concomitant variations are:

- It uses dichotomous variables. Any information, quantitative or qualitative, objective or subjective can be transformed into one or more dichotomous variables. As a result it permits to keep track of all the elements of the phenomena.
- It allows the study of many variables at the same time, thereby it can take into account the complexity of the phenomena. In contrast the experimental method can study only one or a limited number of variables at a time, thereby it produces knowledge which is detached from the context and the complexity of natural phenomena.
- It allows to control for intervening and spurious variables, and this is done after and not before. Therefore, it does not

always need controlled environments such as a laboratory and it is possible to use natural contexts.

— With subjective answers people often respond using masks. For example, even when we feel unhappy, lonely, depressed, usually we try to give an image of ourselves (a mask) which is positive. With the experimental method masks constitute a problem which is insurmountable and which is solved only by removing qualitative and subjective information from the analyses. On the contrary, the methodology of concomitant variations can handle correctly responses which are masked.

This happens because a property of masks is that they affect not only one variable, but all those which are correlated. For example, if a person responds by saying no to "*I feel depressed*," when he is depressed, he will also say no to "*I feel unhappy*," when he is unhappy. The concomitance between depression and unhappiness remains unchanged, because both responses have moved in the same direction and continue to remain correlated.

	Depressed	Not Depressed	Total
Unhappy	15	3	18
Happy	2	*180*	182
Total	17	183	200

Concomitances between masked answers

This table shows that the two modalities, "*I feel happy*" and "*I do not feel depressed*", are concomitant.

When using psychological tests, which produce "objective" measurements of depression and happiness which are not distorted by the effect of masks, answers shift from the positive to the negative side. But the result remains unchanged:

	Depressed	Not Depressed	Total
Unhappy	*158*	10	168
Happy	2	30	32
Total	160	40	200

Concomitances obtained when using "objective" information

Results continue to show the concomitance between the variables depression and unhappiness.

This means that if a concomitance exists it will show also when responses are masked, since masks are applied in a coherent way to all those variables which are correlated. This is a fundamental point, as the problem of masks is ubiquitous in psychological, social and economic sciences. The methodology of concomitant variations solves this problem and allows in this way to widen science to subjective and qualitative data and allows the methodology of concomitant variations to use direct questions, such as: *"do you feel depressed?"*

- Statistics

When using the methodology of concomitant variations, the first thing we have to do is to define which is the "statistical unit." Statistical units allow the study of concomitances among variables and the choice of the statistical unit is strictly related to the aim of the research. Units can be persons, animals, plants, manufactured items, organizations.

With the methodology of differences units are in a one-to-one correspondence with the data values, whereas with the methodology of concomitant variations there is a one-to-many correspondence, since unlimited data values can be collected for each unit.

Sample requirements differ according to the methodology and aim:

- When the aim is to make inferences about the population from the sample, the sample must be representative of the population. This is usually achieved by random sampling.
- When the aim is to study differences among the experimental and the control group the sample must be homogeneous. This is usually achieved by randomly distributing the units across the experimental and control group. If the

aim is to assess the effect of a new drug against a placebo drug, then the patients should be allocated to either the drug group (experimental) or to the placebo group (control) using randomization. Randomization reduces biases by equally distributing factors that have not been explicitly accounted for. When randomization does not allow for the formation of homogeneous groups, the alternative is to use laboratory animals, purposely bred in order to guarantee homogeneity. Laboratory animals are euthanized after being used once, since their use in one experiment makes them different and unsuitable for other experiments.

— When the aim is to study concomitant variations among variables, the sample must be heterogeneous. If the aim is to study which factors cause drug addiction, we will include in the sample subjects with different levels of drug addiction. The definition of the sample is therefore strictly related to the aim. With the methodology of concomitant variations it is important to keep track of all the possible intervening variables, and check later for intervening and spurious relations.

The methodology of differences assesses effects by:

- comparing the difference between mean values of the experimental and control groups with the variability of the values in the sample;
- or by comparing the variance between groups with the variance within groups.

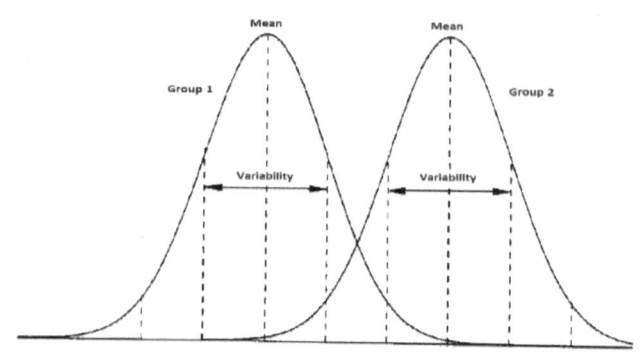

Comparison of mean and variability of two groups

Initial similarity between groups is a fundamental requirement, without which it is impossible to state that the difference observed between the experimental and the control group is a consequence of the cause/treatment. But, in clinical trials the variability of subjects can be so great that even increasing the sample size does not lead to statistical significant results.

When this is the case laboratory animals are used. Laboratory animals are all very similar and decrease the variability of the sample, allowing in this way small differences to become statistically significant.

There is now mounting evidence that animal experimentation constitutes an artifact.[55] The reason is very simple. Statistical significance is stronger when the variability is smaller. Consequently, when the effect size is small, the only way to obtain statistically

---

[55] In experimental science, the expression 'artifact' is used to refer to experimental results which are not manifestations of the natural phenomena under investigation, but are due to the particular experimental arrangement, and hence indirectly to human agency.

significant results is to reduce the variability of the sample. When using animals, which are all very similar, the variability of the sample tends to be null, and consequently also insignificant differences become statistically significant. In other words, animals are too similar and differences that have no actual value become significant. Furthermore, one of the fundamental rules in science is to use samples that are representative of the population to which results will be generalized. It is obvious that laboratory animals are not representative of humans and that the effects observed using laboratory animals are difficult to generalize to humans.

Finally, the methodology of differences uses parametric statistical techniques, which require data distributed according to the Gaussian curve. This condition is usually not met, nevertheless researchers go on and interpret results.

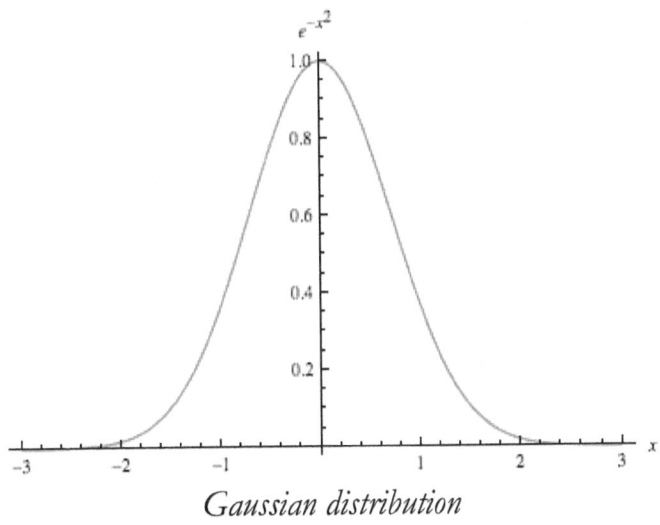

e^{-x^2}

Gaussian distribution

Concomitances require variability: heterogeneous samples, where variability is maximized. The methodology of differences requires homogeneous samples, whereas the methodology of concomitant variations requires heterogeneous samples.

For example, with the methodology of concomitant variations, in a study that aims to compare the growth of 5 different types of crop in 5 different types of field, all the combinations will be considered (5! = 120 possible combinations) and at least 30 measurements will be taken for each combination. Since the aim is to compare growth rates, the statistical unit will be the height of the crop after a fixed interval of days (or a similar type of measurement). For each measurement an array of information will be traced, such as the type of field and the type of crop, secondly information that we think can be related to the growth of crop. At the end we will have 3600 records (30 measurement x 120 combinations), each with data on the growth rate and an array of other information.

When answers tend to concentrate in one modality, wider measuring scales are needed. For example, when we ask *"Do you feel depressed?"* yes/no, most people answer no and this little variability limits the possibility of studying concomitances. In order to restore variability it is necessary to use wider scales, such as *"How much do you feel depressed?"* *0,1,2,3,4,5,6,7,8,9,10.* Most answers will concentrate in the low values, 0 to 3, and the median cut-off point will probably be between the values 1 and 2. The aim of the methodology of concomitant variations is to study relations maximizing the variability.

Usually at least 100 units (ie subject/records/forms) are required. But, in many clinical studies only one subject is available. When this is the case, measurements can be repeated in different moments, trying to maximize the variability. For example, if we want to study what is concomitant to our headaches, we keep track at regular intervals of all what we think might be related to this situation. For example, each evening we fill a form in which we provide a subjective measurement of the headache, plus what we ate, what we watched on TV, our feelings, etc. When a sufficient number of forms (possibly more than 100) is filled we can process them.

Data can be collected in various ways: nominal, ordinal, interval and ratio.

- *Nominal* or categorical data, are made of mutually exclusive modalities. For example: marital status, nationality.
- *Ordinal* data, are variables where the order matters but not the difference between values. For example, if we ask patients to

express the amount of pain they are feeling on a scale of 0 to 10. A score of 7 means more pain than a score of 5, and 5 is more than a score of 3. But the difference between 7 and 5 may not be the same as that between 5 and 3. The values simply express an order, a progression.

- *Interval* data, are variables where the difference between two values is meaningful. For example the difference between 1 meter and 2 meters is the same difference as between 3 and 4 meters. That is, numbers are spaced always by the same measuring unit.

- *Ratio* data have all the properties of interval variables, but have also a clear definition of the zero value. Variables like height, weight, enzyme activity are ratio variables. Temperature, expressed in Fahrenheit or Celsius, is not a ratio variable. A temperature of zero degrees on either of those scales does not mean no temperature. Kelvin degrees correspond instead to a ratio variable since zero degrees Kelvin really correspond to no temperature. When working with ratio variables, but not interval variables, it is possible to use divisions. A weight of 4 grams is twice a weight of 2 grams, because weight is a ratio variable. A temperature of 100 degrees Celsius is not twice as hot as 50 degrees Celsius, because temperatures in Celsius are not a ratio variable. The Celsius scale is an interval variable, whereas the Kelvin scale starts from absolute zero and allows for ratios.

The mathematical operations which can be performed are:

— in the case of nominal/categorical variables the value is a modality of a list, for example Italy France, Germany. With these variables it is possible only to count the occurrences of the modalities.

— In ordinal variables the value is a sequence: First, Second, Third; Elementary education, High School,

University. It is possible to divide the sequence into high and low, for example high education, low education, or treat each value as a modality (nominal variable). For example, it is possible to count how many people have reached secondary or higher education. It is possible to find which is the level of education attained at least, for example, by 50% of the population. There is an order, a progression, which can be used to create new categories (e.g. low education and high education) or to order the population. Ordinal variables allow for counting and sorting.

— Interval variables allow to calculate average values and variabilities since they permit the use of additions and subtractions.

— Ratio variables use the absolute zero value and allow to use divisions and multiplications.

Data can be transformed in one or more dichotomous variables.

- In the case of nominal variables, the single modality (e.g. single province, nationality, color) can be translated into a dichotomous variable. For example, Italy becomes the Italy dichotomous variable for which the answers can only be: yes or no.
- Ordinal variables follow a progression. These variables can be treated in the same way as the nominal variables by translating each modality in a dichotomous variable, but it is also possible to translate the information in the form high/low. It is important to note that there is no objective criterion for defining when modalities are considered high or low. For example, in a study concerning university professors the lowest degree of education might correspond to the highest degree in another study which considers the poor population of developing countries. The division of an ordinal variable into a dichotomous variable, must always take

into account the context and purpose of the study. In the event that no criterion suggests how to divide between high and low the cut-off point is chosen by balancing the two groups. This is done using the median value.

– When dealing with interval or ratio variables cut-off values, that mark the transition from low to high values, are generally used. The aim of the researcher and the purpose of data analysis is usually to identify these cut-off values. It happens frequently that the same variable can be translated into multiple dichotomous variables in order to test which cut-off value best allows to identify a critical value, i.e. a value that indicates the transition from one state to another.

Data is the row material, but not all data is suitable for concomitant variations analyses; only data which can be transformed in the dichotomous form and is gathered in a systematic way can be used. Information which cannot be coded or transformed in the dichotomous form is of little use.

In the late 19th century, Charles Sanders Peirce in *"How to Make Our Ideas Clear"*[56] placed induction and deduction in a complementary rather than competitive context. Secondly, and of more direct importance to scientific method, Peirce put forth the basic schema for hypothesis-testing that continues to prevail today. Peirce examined and articulated the fundamental modes of reasoning that play a role in scientific inquiry, the processes that are currently known as abductive, deductive, and inductive inference:

1. During the *inductive* phase we consciously review the know-how and unsolved problems.
2. During the *abductive* phase unconscious processes take place and lead to intuition which highlights new hypotheses and solutions.

---

[56] Peirce C.S. (1878), How to Make Our Ideas Clear, www.amazon.it/dp/B004S7A74K

3. During the *deduction* phase hypotheses are translated into items.

4. During the *validation* phase data is gathered and hypotheses and solutions are tested.

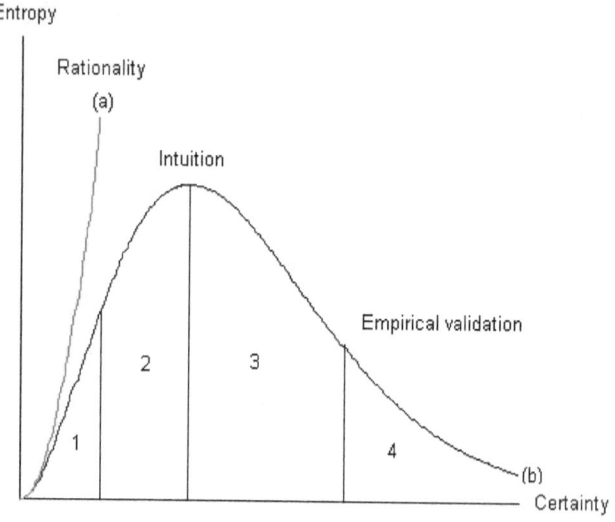

Phases of the process of discovery

One of the most delicate phases is when we translate hypotheses into items (phase 3).

Hypotheses always state a concomitance between two or more variables. In order to test these concomitances it is required to gather data separately. For example if the hypothesis is that loneliness causes anxiety it is wrong to ask: *Loneliness causes anxiety?* because the concomitance between loneliness and anxiety is already given in the item and data analysis will not be able to tell if this concomitance exists.

In order to study the concomitance between loneliness and anxiety it is necessary to formulate two different items: *Do you feel lonely? Do you experience anxiety?*

Data analysis will tell if these two items (loneliness and anxiety) vary in a concomitant way and are related. It is also important to ask information in a clear and direct way, avoiding negative forms. Each item should contain only one information.

For example the following item is incorrect since it combines State Aid (Yes/No) with Family type (one parent family, two parents family):

Did the family receive State Aid?
   ☐ Yes, No,
   ☐ It is a one parent family,
   ☐ It is a two parents family

The correct formulation is:

Did the family receive State Aid? Yes, No

Family type: One parent, Two parents

Each item (i.e. each variable) must be relative only to one type of information. During data analysis information will be combined and concomitances will be studied.

Items can be divided into key items, explicative and structure items:

- *key items* are all those variables which describe the topic under investigation, for example if the study is relative to cancer, key variables will be relative to cancer;
- *explicative items* are all those variables which might be correlated (linked) to the key variables, for example in the case of cancer it could be the environment, stress, food, and so on;
- *structure items* are variables such as age, sex, education, profession; variables which are

usually used to describe the sample of the study and the context.

In order to choose relevant explicative variables, it can be useful to ask the help of experts who have a good knowledge of the subject. It is also useful to compare different hypotheses. Scientific research is a process of continuous evolution of knowledge which requires the disposition to revisit, change and eventually abandon our beliefs.

Designing a form can be divided in the following steps:

— declare which is the aim of the study (*key variables*).

— list all those variables (*explicative variables*) which might be correlated (concomitant) to the key variables. It is very important to keep track of the hypotheses, in this way the interpretation of the results will be straightforward, otherwise it is easy to fall in the trap of paying too much attention to secondary information and produce interpretations which are totally irrelevant and of little scientific value. It is always a good habit to use more items for the same information (redundancy).

— prepare the form (questionnaire, observation grid, ...) and test it in order to assess if it works well or if it can be improved and optimized. It is necessary to continue testing the form until it reaches a standard which we consider acceptable.

Parametric statistical tests are based on the assumption that the variables data in the population are distributed according to the normal (Gaussian) distribution, which in probability theory is a continuous distribution, a function, which allows to calculate the probability that any real observation will fall between any two limits.

On the contrary, nonparametric methods make no assumptions about the distribution of data. Their applicability is much wider than the corresponding parametric methods and, due to the reliance on fewer assumptions, are more robust and simple. Even when the use of parametric methods is justified, nonparametric methods are easier to use and more reliable. Because of their simplicity, results leave less room for improper use and misunderstanding.

In the 1960s Simon Shnoll and co-workers were probably the first scientists to show that the assumption of the normal distribution is only mathematical, and that in life sciences and also in physics it is false.

In a review of studies performed over more than forty years, Shnoll[57] shows the non-randomness of the fine structure of the distributions of measurements, starting from biological objects and moving into the purely physical domain. The implication is huge: tests based on the assumption of normal random distributions, such as those in the field of parametric statistics, are fundamentally biased and produce results

---

[57] Shnoll SE, Kolombet VA, Pozharskii EV, Zenchenko TA, Zvereva IM and AA Konradov, Realization of discrete states during fluctuations in macroscopic processes, Physics – Uspekhi 162(10), 1998, pp.1129–1140.
http://ufn.ioc.ac.ru/abstracts/abst98/abst9810.html#d

which are often unstable and difficult to reproduce.

The methodology of concomitant variations uses nonparametric statistics, among which the Chi Square (χ^2) is today one of the most widely used statistical indexes. χ^2 calculates the differences between observed frequencies and expected frequencies. In the absence of concomitances χ^2 is equal to 0, whereas in the case of maximum concomitance it is equal to the size of the sample.

The comparison with the χ^2 probability distributions allows to know the statistical significance of the concomitance. Statistical significance indicates the risk which is accepted when we state the existence of the relation. Conventionally concomitances are taken in consideration when the risk is below 1%.

With dichotomous variables concomitances can be accepted with a risk lower than 1%, with χ^2 values greater or equal to 6.635.

When using the methodology of

concomitant variations all variables are translated into the dichotomous form. Crossing two dichotomous variables produces a 2x2 table. If we take, for example, the following variables **A** and **B**:

		A	
B	Yes	No	**Total**
Yes	18,340	3,241	**21,581**
No	5,118	29,336	**34,454**
Total:	**23,458**	**32,577**	**56,035**

the χ^2 value is obtained by comparing the observed frequencies and the expected frequencies.

Expected frequencies are calculated by dividing the product of the total values of row and column by the general total. For the expected frequency of the first cell (Yes / Yes) is:

$$21,581 \times 23,458/56,035 = 9,034$$

Following this procedure for all the cells of the table we have the following expected frequencies table:

B	A Yes	No	Total
Yes	9,034	12,547	21,581
No	14,424	20,030	34,454
Total:	23,458	32,577	56,035

The Chi Square formula is the following:

$$Chi\ Square = \sum \frac{(f_o - f_e)^2}{f_e}$$

where f_o indicates observed frequencies and f_e expected frequencies

For each cell we calculate the square of the difference between observed frequencies and expected frequencies divided by expected frequencies and we sum the results together.

In this example we obtain a Chi Square value of 26,813, well above the value 6.635 from which the statistical significance of 1% starts.

Since the maximum value of χ^2 varies depending on the number of cases, it is useful to standardize it between 0 and 1. This transformation is known as the *rPhi* and is obtained as the square root of the value of χ^2 divided by the sample size and behaves similarly to Pearson's correlation index.

Correlations/concomitances can be of two types: direct or inverse. If the correlation is directed the two dichotomous variables are concomitantly true or false, whereas if the correlation is inverse one variable is true when the other is false.

Inverse correlations have negative sign (-), whereas direct correlations are shown without sign.

- Software

The Sintropia-DS software was developed in order to make the methodology of concomitant variations available. A complete description is available in the help sections of the software, or in the dedicated 2005 issue of the Syntropy Journal.[58]

The first version of Sintropia-DS dates

---

[58] www.sintropia.it/journal

back to 1982, it was distributed with the name DataStat, and extensively used in the Department of Statistics of the University of Rome. Sintropia-DS merges database and statistical analyses (this is the reason of the extension DS: database and statistics).

In order to install Sintropia-DS in your computer: download the zip file from www.sintropia.it/sintropia.ds.zip, copy the folder "Sintropia.DS" from the zip file in the root disk "C:", and find the Sintropia application in the folder Sintropia.DS.

Since this version of the software dates back to 2005 and was developed for Windows-XP, recent version of Windows require that you allow the use of the program.

SYNTROPY
AND
QUANTUM MECHANICS

At the end of the 19th century Lord Rayleigh and Sir James Jeans extended the equipartition theorem of classical statistical mechanics to an ideal black body at thermal equilibrium and were faced with a fundamental paradox.

According to the equipartition theorem a black body at thermal equilibrium (which in physics is the best possible emitter of thermal radiation) will emit radiation with infinite power as it would all concentrate in the ultraviolet wavelength.

This prediction was named the *ultraviolet catastrophe*, but fortunately it was not observed in nature.

The paradox was solved on 14 December 1900 when Max Planck presented a paper, at the German Physical Society, according to which energy is quantized.

Planck assumed that energy does not grow or diminish in a continuous way, but according to multiples of a basic quantum, which Planck defined as the frequency of the body (v) and a basic constant which is now known to be equal to $6,6262 \cdot 10^{-34}$ joule $\cdot$ seconds and which is now named Planck's constant.

Planck described thermal radiations as made of packets (quantum), some small and others larger according to the frequency of the body. Below the quantum level, thermal radiation disappeared, avoiding in this way the formation of infinite peaks of radiation at the ultraviolet wavelength and solving in this way the paradox of the ultraviolet catastrophe.

December 14, 1900, is now remembered as the starting date of quantum mechanics.

Quantum theory was further confirmed by Einstein with the study of the photoelectric effect.

When light or electromagnetic radiation reach a metal, electrons are emitted, this is named the photoelectric effect. The electrons of the photoelectric effect can be measured, and these measurements show that:

- until a specific threshold is reached the metal does not emit any electrons;
- above the specific threshold electrons are emitted, and their energy remains constant;
- the energy of the electrons increases only if the frequency of light is raised.

Classical light theory was not able to justify this behavior:

- Why does the intensity of light not increase the energy of the electron emitted by the metal?
- Why does the frequency affect the energy of the electrons?
- Why are electrons not emitted below a specific threshold?

In 1905, Einstein answered these questions using Planck's constant and suggesting that light, previously considered an electromagnetic wave, could be described as quantum packets of energy, particles which are now called photons.

Einstein's interpretation of the photoelectric effect played a key role in the development of quantum mechanics, as it treated light as particles, instead of waves, opening the way to the duality wave/particles.

The experimental proof of Einstein's interpretation was given in 1915 by Robert Millikan who, ironically, had been trying, for 10 years, to prove that Einstein's interpretation was wrong. In his experiments Millikan discovered that all the alternative theories did not pass the experimental test, whereas only Einstein's interpretation was shown to be correct.

Several years later Millikan commented:

"I spent ten years of my life testing that 1905 equation of Einstein's and contrary to all my expectations I was compelled in 1915 to assert its unambiguous experimental verification in spite of it unreasonableness since it seemed to violate everything that we knew about the interference of light."

Planck himself remained skeptical of his own discovery failing to answer the question "*why quantum*?" This question has not yet received an answer and remains one of the fundamental mysteries of quantum mechanics.

Syntropy suggests that atoms vibrate between diverging and converging phases. In the diverging phase, atoms can emit a packet (quantum) of energy, whereas during the converging phase they can absorb a quantum. In the diverging phase entropic energy is accessible, whereas in the converging phase syntropic energy is accessible.

This vibrating interpretation of the atom can answer several questions. For example, according to the second law of thermodynamics, particles (such as the electron) should rapidly lose their kinetic charge and fall towards the center of the atom. This does not happen.

Syntropy suggests that atoms vibrate in infinite cycles of expansion and contraction, in which the effect of entropy is counterbalanced by syntropy during the converging phase.

In this interpretation the wave/particle duality is the manifestation of the duality: causality/retrocausality, entropy/syntropy. Where causation is deterministic and retrocausality is probabilistic. Two types of causality united by the same energy and coexisting in every manifestation of matter.

But, the negative energy time solution was considered to be impossible since it introduces retrocausality and the possibility of perpetual motion in physics (perpetual motion is though observed in atoms!).

In order to avoid retrocausality Einstein argued that the momentum is negligible, since the motion of bodies is practically nil, when compared to the speed of light. When the momentum is set equal to zero ($p=0$), the energy/momentum/mass equation simplifies into the famous $E=mc^2$, which always has positive solution, without any reference to the direction of time.

In 1924 Wolfgang Pauli, one of the pioneers of quantum mechanics, discovered that electrons have a spin, a momentum which nears the speed of light. As a result it was necessary to combine quantum mechanics and special relativity, using the $E^2=m^2c^4+p^2c^2$ formula and not the simplified $E=mc^2$.

In 1925 the physicists Oskar Klein and Walter Gordon formulated the first equation that combined quantum mechanics and special relativity and found themselves with two solutions: one that describes matter and energy that propagate forward in time, and the other describing matter and energy that propagate backward in time (now known as antimatter).

In 1926 Erwin Schrödinger removed the energy/momentum/mass equation from Klein and Gordon's equation obtaining in this way his famous wave function (Ψ).

In 1927, Klein and Gordon formulated again their equation as a combination of Schrödinger's wave function and the energy/momentum/mass equation.

The equation of Klein and Gordon manages to explain the mysteries of quantum mechanics, such as the duality wave/particle which would result from the duality causality/retrocausality. However Niels Bohr and Werner Heisenberg considered retrocausality unacceptable. Starting from the Schrödinger equation, which treats time in the classic time-forward way, they suggested what is now known as the Copenhagen interpretation of quantum mechanics, which states that matter propagates as a wave and only when it is observed the wave collapses into a particle. But the act of observing is an act of consciousness. In this way Bohr and Heisenberg gave to consciousness the power to create reality. This interpretation fitted the Nazi ideology, which stated that men are endowed with powers of creation.

When Erwin Schrödinger discovered how Heisenberg and Bohr had used his equation, with ideological and mystical implications, he commented: *"I don't like it, and I am sorry I ever had anything to do with it."*

In 1928, Paul Dirac, tried to solve the dispute by applying the $E^2 = m^2c^4 + p^2c^2$ equation to the electron. To his disappointment, he obtained two solutions: the electron and the neg-electron, where the electron moves forward in time and the neg-electron backward in time.

The neg-electron caused emotional distress. For example Heisenberg wrote to Pauli: *"The saddest chapter of modern physics is and remains the Dirac theory ... I regard the Dirac theory as learned trash which no one can take seriously."*

In 1931, in an effort to remove the unwanted retrocausal solution, Dirac used Pauli's principle, according to which two electrons cannot share the same state, to suggest that all states of negative energy are occupied, thereby forbidding any interaction between forward-in-time and backward-in-time states of matter. On this assumption of an ocean of negative energy, called the Dirac sea, the more recent "Standard Model" of physics continues to be based.

However, in 1932 Carl Anderson discovered neg-electrons in cosmic radiations and named them positrons, thus paving the way for the study of antimatter.

The scientific debate between special relativity and quantum mechanics was soon poisoned by political passions. In April 1933 Einstein learned that the new German government had passed a law excluding Jews from holding any official positions, including teaching at universities. A month later, the episode of the burning of books by the Nazis occurred, with Einstein's works being among those burnt, and Nazi's propaganda minister Joseph Goebbels proclaimed, *"Jewish intellectualism is dead."* Einstein's name was on a list of assassination targets, with a *"$5,000 bounty on his head"* and one German magazine included him in a list of enemies of the German regime with the phrase, *"not yet hanged."* Einstein's treatises were burned, his suburban villa in Berlin was raided, and his furniture, books, bank account and even his violin were seized. Hitler's ideological convictions about Jewish science had received support from the book *"100 Authors against Einstein."*[59] The theory of relativity was stigmatized as Jewish science, deliriums of a crazy Jew whereas the Copenhagen interpretation was imposed.

---

[59] Israel H (1931), Ruckhaber E e Weinmann R, Hundert Authoren gegen Einstein, Voigtlanders, Peipzig, 1931.

In the Copenhagen Interpretation, the collapse of the wave function (wave collapsing into particles) occurs at the same moment in all the points of the wave. This implies an instantaneous propagation of information that violates the limit of the speed of light that Einstein considered the upper limit for the propagation of information and causality.

Einstein considered causality always local and speeds had always to be lower or equal to that of light, but never faster.

Starting from these assumptions Einstein rejected the idea that the information of the collapse of the wave function could propagate instantaneously and, in 1934, he formulated the EPR paradox which was named from the initials of the persons who formulate it (Einstein-Podolsky-Rosen).

EPR started from Pauli's discovery that electrons have a spin and that the same orbit can be shared by only two electrons with opposite spins (the Pauli exclusion principle). The Copenhagen Interpretation concludes that electron pairs which shared the same orbital remain correlated (entangled) showing always opposite spins, regardless of their distance, thus violating the limit of the speed of light in the propagation of information.

The EPR paradox remained unanswered for more than 50 years and was considered as a thought experiment, in order to demonstrate the absurdity of the Copenhagen Interpretation, raising a logical contradiction.

No one expected that the EPR experiments could be carried out, however, in 1952 David Bohm suggested to replace electrons with photons, and in 1964 John Bell showed that this change opened the way to the experiment.

However, at that time, not even Bell believed that the experiment could actually be done. But scientists accepted the challenge and in 1982 the team of Alain Aspect, published the results that show that Einstein was wrong.[60]

The quantum property measured by Aspect is the polarization of the photon, which can be imagined as an arrow which points upwards or

[60] Aspect A (1932), Experimental Realization of Einstein-Podolsky-Rosen-Bohm, Gedankenexperiment, Physical Review Letters, vol. 49, 91, 1982.

downwards.

We can stimulate an atom to produce two-photon simultaneously, which are sent in two different directions. The polarizations of the two photons must be opposite: if the arrow in the first one goes up, the other must go down. Each photon leaves with a well-defined polarization, and the coupled photon with the opposite polarization. Both retain their polarization in their journey through space.

The Copenhagen Interpretation states that any quantum entity with this dual possibility exists in a superposition of states, until its polarization is not measured and the wave function collapses. Only after the wave function collapses the counterpart of the photon that is measured must show the opposite arrow direction. At the precise moment in which the measurement of the photon is performed, the collapse of the wave forces photon B (which could, in principle, be on the far side of the universe) into the opposite state. The instantaneous response of photon B to what happens to photon A is what Einstein called the "*spooky action at a distance.*"

The experiment made by Aspect measured the polarization according to an angle, which can be varied, with respect to upward and downward arrows. The probability that a photon with a certain polarization will pass through a filter arranged with a certain angle depends on its polarization and the angle between the polarization and the filter. In a non-local reality changing the angle with which the polarization of the photon A is measured will necessarily alter the probability that the photon B passes through a polarizing filter arranged at a different angle. In addition, the experiment not only considers two photons, but entire beams of photons, or series of related pairs whizzing through the apparatus one after the other.

Bell had shown that if Einstein was right the number of photons that go through the B polarizing filter had to be lower than that which passes through filter A. This takes the name of Bell inequality. However, Aspect's experiment proved the opposite, that the first value (A) is always lower to the second value

(B). To put it in other words, Bell inequality is violated, and the common sense embodied by Einstein lost the challenge.

Although Aspect's experiment was motivated precisely by quantum theory, Bell's theorem has much broader implications and the combination of Bell's theorem and the experimental results reveals a fundamental truth of the universe, that there are correlations which take place instantly, regardless of the distance between objects, and that signals seem to be able to travel at speeds exceeding that of light.

As a result of the EPR paradox and the results of Aspect's on non-locality and entanglement, quantum mechanics and special relativity are generally considered to be incompatible even if both are accurate in predicting the results of the experiments.

The conflict between quantum mechanics and special relativity unravels when we accept the possibility of retrocausality: effects that can propagate backwards in time, and that can occur instantaneously in space, and travel at speeds which exceed that of light.

In his book *"The Road to Reality"* Roger Penrose underlines that usually physicists tend to reject as "unphysical" any solution which contradicts classical causality, according to which causes always precede effects. Any solution which makes it possible to send a signal backward-in-time is usually rejected.

Even if Penrose chose to reject the negative time solution of the energy equation, he states that this refusal is a consequence of a subjective choice, towards which other physicists have different opinions.

Penrose dedicates nearly 200 pages of his book to the paradox of the negative time solution. According to Penrose it is important that the value of E is always positive because negative values of E lead to catastrophic instabilities in the Standard Model of sub-atomic physics.

"Unfortunately in relativistic particles both solutions of the equation need to be considered as a possibility, even a non-physical negative energy has to be considered as a possibility. This does not happen in

non-relativistic particles. In this last case, the quantity is always defined as positive, and the embarrassing negative solution does not appear."[61]

Penrose adds that the relativistic expression of Schrödinger's equation (ie the equation of Klein Gordon) does not offer a clear procedure in order to exclude the backward-in-time solution of the square roots.

In the case of a single free particle (or a system of non-interacting particles), this does not lead to a serious difficulty, because we can restrict our attention to overlapping plane wave solutions of positive energy of Schrödinger's equation. However, this is no longer the case when there are interactions; even for a single relativistic particle charge in an electromagnetic field, the wave function cannot, in general, maintain the positive time solution. This creates a conflict with the law of cause and effect as it introduces the possibility of retrocausality, of causes that retroact from the future.

---

[61] Penrose R (2005), *The road to reality: A Complete Guide to the Laws of the Universe*, Knopf Doubleday Publishing Group, 2005

Despite the fact that the official position is to reject retrocausality, a growing number of physicists is working on this possibility.

Richard Feynman's diagrams of electron-positron annihilation offer an example.

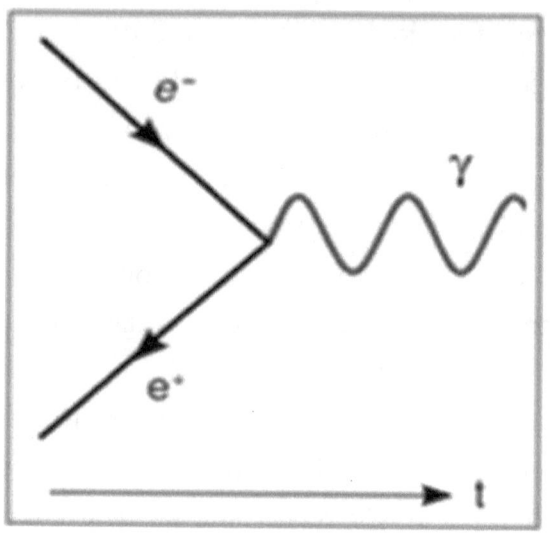

In the diagram arrows to the right represent electrons, arrows to the left represent positrons, wavy lines photons.

According to these diagrams, electrons do not annihilate when they get in contact with positrons, but they release energy since they change their time direction becoming positrons and starting to move backward-in-time.

When Feynman diagrams are interpreted they necessarily imply the existence of retrocausality.

John Archibald Wheeler and Richard Feynman used the backward-in-time energy solution of the wave equation, the *"advanced waves"* solution, to solve Maxwell's equations.

Feynman has also used the concept of retrocausality to produce a model of positrons which reinterprets Dirac's hypothesis of the sea of negative energy occupying all possible states. In this model, electrons which move backward-in-time acquire positive charges.

In 1986 John Cramer, physicist at Washington State University, presented the Transactional Interpretation of quantum mechanics.[62] The outcome of the experiments remains exactly the same as those of the other quantum interpretations, but what characterizes this interpretation is the different perspective on what is happening, that many find easier and simpler. In this interpretation

---

[62] Cramer JG (1986), The Transactional Interpretation of Quantum Mechanics, Reviews of Modern Physics, 1986, 58: 647-688.

the formalism of quantum mechanics is the same, but the difference is how this formalism is interpreted.

Cramer was inspired by the absorber-emitter theory developed by Wheeler and Feynman which used the dual solution of Maxwell's equation. As it is well known also the generalization of Schrödinger's wave equation into a relativistic invariant equation (Klein-Gordon's equation) has two solutions, one positive, which describes waves which propagate forward in time, and one negative, which describes waves which propagate backward-in-time.

This dual solution allows to explain in a simple way the dual nature of matter (particles and waves), non-locality and all the other mysteries of quantum mechanics and permits to unite quantum mechanics with special relativity.

The transactional interpretation requires that waves can really travel backward-in-time. This assertion is counterintuitive, as we are accustomed to the fact that causes precede effects. It is important to underline that the transactional interpretation takes into account special relativity, which describes time as a dimension of space, in a way which is totally different from our usual way of thinking.

The Copenhagen interpretation, instead, treats time in the classical Newtonian approach, and this leads to the use of consciousness in a mystical way.

The probabilistic equation developed by Max Born in 1926 contains an explicit reference to the nature of time and to the two possible solutions of the advanced and delayed waves. Since 1926, every time physicists have used Schrödinger's equation in order to calculate quantum probabilities, they have considered the advanced waves solution without even realizing it.

Cramer's mathematics is exactly the same of the Copenhagen interpretation. The difference lies solely in the interpretation. Cramer's interpretation solves all the mysteries of quantum physics, making it also compatible with the requirements of special relativity. This miracle is achieved, however, at the price that the quantum wave can actually travel back in time. At first glance, this is in sharp contrast with common logic, which tells us that causes must always precede effects, but the way in which the transactional interpretation considers time differs from common logic, since the transactional interpretation explicitly includes the effects of the theory of relativity.

The Copenhagen interpretation, instead, treats time in the traditional Newtonian way, and this is the cause of the inconsistencies and paradoxes which are observed in the experiments.

Yoichiro Nambu (2008 Nobel Prize for physics) has applied Feynman's model to the processes of annihilation of particle-antiparticle couples, arriving at the conclusion that it is not a process of annihilation or creation of couples of particles and antiparticles, but simply a change of the time direction of particles, from the past to the future or from the future to the past.[63]

---

[63] Nambu Y. (1950) The Use of the Proper Time in Quantum Electrodynamics, Progress in Theoretical Physics (5).

In 1977 Costa de Beauregard used the concept of retrocausality to explain quantum entanglement.[64]

The idea that the arrow of time can be reversed is very recent. Until the XIX century, time was considered to be irreversible, a sequence of absolute moments. Only with the introduction of special relativity the concept of retrocausality started entering the scientific world.

In 1954 the philosopher Michael Dummett showed that there is no philosophical contradiction in the idea that effects can precede causes.[65]

In 2006 AIP (American Institute of Physics, 2006) organized a conference in San Diego California titled "Frontiers of Time: Retrocausation – Experimental and Theory."[66]

In November 2010, President Barack

[64] De Beauregard C (1977), Time Symmetry and the Einstein Paradox, Il Nuovo Cimento, 1977, 42B.
[65] Dummett M (1954), Can an Effect Precede its Cause, Proceedings of the Aristotelian Society, 1954, Supp. 28.
[66] American Institute of Physics (2006), Frontiers of Time. Retrocausation – Experimental and Theory, AIP Conference Proceedings, San Diego California, 20-22- June 2006.

Obama awarded the physicist Yakir Aharonov the National Medal of Science for the experimental studies which show that the present is a result of causes which flow from the past as well as from the future. These results suggest a radical reinterpretation of time and causality.[67]

Einsteins' Special relativity started a new description of reality: on the one hand energy and matter that propagate from the past to the future, on the other energy and matter that propagate backward in time from the future to the past.

Einstein used the term Übercausalität (supercausality) to describe this new model of time that combines causality and retrocausality.

In the paper "*A novel interpretation of the Klein-Gordon equation*," Wharton concludes that:

"*It is obvious that quantum mechanics is counter-intuitive, but it must be counter-intuitive for a reason — some human intuition that fundamentally contradicts some physical principle. One example of this would be*

---

[67] Aharonov Y (2005), Quantum Paradoxes, Whiley-VCH, Berlin, 2005.

the well-known conflict between our direct experience of time and the more symmetric treatment of time in fundamental physics. If the counter-intuitive aspects of quantum mechanics could be explained via classical fields symmetrically constrained by both past and future events, then it would be a mistake to reject such a solution based solely on our time-asymmetric intuitions."[68]

In the special issue *"Emergent Quantum Mechanics – David Bohm Centennial Perspectives"* published in *Entropy*, retrocausality is extensively reviewed with a total of 126 references.[69] This shows that the concept of retrocausality is finally entering the field of physics.

In the words of Richard Feynman the wave/particle duality contains the "central mystery" of quantum mechanics:

---

[68] Wharton KB (2009), A novel interpretation of the Klein-Gordon equation, Foundation of Physics, 2009, 40(3): 313-332.
[69] Walleczek J, Grössing G, Pylkkänen P and Hiley B (2019) *Emergent Quantum Mechanics – David Bohm Centennial Perspectives*, www.mdpi.com/books/pdfview/book/1203

"The double slit experiment is a phenomenon which is impossible, absolutely impossible, to explain in any classical way, and which has in it the heart of quantum mechanics."[70]

Richard Feynman considered this experiment so important that he dedicated to it the first chapter of the third volume of his famous *"Lectures on Physics."*

Syntropy and the dual solution of the Klein-Gordon equation predict the duality wave/particle, as the manifestation of causality and retrocausality. Particles are the manifestation of causality, whereas waves are the manifestation of retrocausality (not yet determined and probabilistic).

The Klein-Gordon equation describes reality as a continuous interplay between emitters and absorbers, causality and retrocausality, causes and attractors.

In the absence of one of these two, there

---

[70] Feynman R.P., et al. (2006),The Feynman Lectures on Physics, Addison Wesley. 4-1.

would be no exchange of matter or energy.

If only causality exists, that is the emitting part, a battery would have a single electron-emitting pole. On the contrary, two poles are needed, one that emits and the other one that absorbs. In the absence of this duality, touching only the emitter (-) or the absorber pole (+), there is no flow of electricity.

In the quantum level, this continuous interplay between causality and retrocausality (emitters / absorbers) causes matter to always manifest as waves and particles combined together.

The duality waves/particles supports the supercausal nature of reality with past and future constantly interacting.

EPILOGUE

Generally we tend to overlook the invisible dimension as it is widely believed that it does not exist and that decisions should be based only on facts. This attitude has led people away from insights, inspirations and dreams and has limited decision making only to rational processes that increase entropy.

This has been very useful during the industrial revolution which has shaped Western culture and societies, but it is now dysfunctional.

Teilhard de Chardin noted that:

"Right now, as in Galileo's days, what is most essential (...) is a new way of thinking, tied to a new way of acting."

The signs of extending science to a new supercausal paradigm which takes into account also the invisible side of reality, can be seen a bit everywhere, but are still not welcomed. Teilhard was exiled in China and the Vatican banned the works of Teilhard from all the libraries since they *"offend the Catholic doctrine."*

Fantappiè was censored. The following words of Francesco Severi[71], founder of the National Institute of Higher Mathematics of Rome, well describe this situation:

"About the problem of finality I am very embarrassed to express an opinion on what someone very close to me calls the discovery of scientific finalism. Science ceases to be science when its results do not express causal results. It is possible to speak of finality in science, but only in a metaphysical sense, having no claim to prove anything positive about it. This is because: 1) it is not possible to deduct hypotheses from the fact that life is subject to final causes, 2) pure logic cannot be used as a scientific demonstration, 3) finality cannot be demonstrated using the experimental method, because no experiment can be established, without acting on the causes prior to the effects. Finalism, in short, is in my opinion an act of faith, not an act of science."

---

[71] Francesco Severi was the founder of the National Institute of Higher Mathematics in Rome.

The situation has now changed.

It is now possible to conduct experiments which test and validate Fantappiè's and Teilhard's hypotheses and this will help the transition from the old paradigm, to the new finalistic and syntropic paradigm.